ROCKS AND MINERALS

TASK CARD SERIES

Conceived and
written by
Doris Metcalf
Ron Marson
Illustrated by
Peg Marson

TOPS LEARNING SYSTEMS

10970 S. Mulino Rd.
Canby OR 97013

NOW YOU CAN PHOTOCOPY!

OLD TASK CARD FORMAT

NEW TASK CARD FORMAT

Dear Educator,
Please excuse our transition . . .

TOPS open-ended task card modules are taking on a new look. Task cards that used to come printed 4-up on heavy index card stock, packaged 2 sets to a zip-lock bag, are now printed 2-up at the back of this single book.

Even though our new cards are printed on lighter book stock, even though we haven't included an extra copy, we can now offer you something much better: You have our permission to make as many photocopies of these task cards as you like, as long as you restrict their use to the students you personally teach. This means you now can (1) incorporate task cards into full-sized worksheets, copying the card at the top of the paper and reserving the bottom for student responses. (2) You can copy and collate task card reference booklets, as many as you need for student use. Or (3) you can make laminated copies to display in your classroom, as before.

It will take some time to fully complete this transition. In the interim we will be shipping TOPS modules as a mixture of both old and new formats. Effective immediately (September 1989) this newer, more liberal photocopy permission applies to all task cards, including our older, heavier, 4-up standards!

Happy sciencing,

Ron Marson
author/publisher

ISBN 0-941008-23-1   Printed on Recycled Paper with Soy Ink ♻

CONTENTS

 PART I   **INTRODUCTION**

 PART II   **TEACHING NOTES**

 PART III   **REPRODUCIBLE STUDENT TASK CARDS**

A TOPS Model for Effective Science Teaching...

If science were only a set of explanations and a collection of facts, you could teach it with blackboard and chalk. You could assign students to read chapters and answer the questions that followed. Good students would take notes, read the text, turn in assignments, then give you all this information back again on a final exam. Science is traditionally taught in this manner. Everybody learns the same body of information at the same time. Class togetherness is preserved.

But science is more than this.

Science is also process — a dynamic interaction of rational inquiry and creative play. Scientists probe, poke, handle, observe, question, think up theories, test ideas, jump to conclusions, make mistakes, revise, synthesize, communicate, disagree and discover. Students can understand science as process only if they are free to think and act like scientists, in a classroom that recognizes and honors individual differences.

Science is *both* a traditional body of knowledge *and* an individualized process of creative inquiry. Science as process cannot ignore tradition. We stand on the shoulders of those who have gone before. If each generation reinvents the wheel, there is no time to discover the stars. Nor can traditional science continue to evolve and redefine itself without process. Science without this cutting edge of discovery is a static, dead thing.

Here is a teaching model that combines the best of both elements into one integrated whole. It is only a model. Like any scientific theory, it must give way over time to new and better ideas. We challenge you to incorporate this TOPS model into your own teaching practice. Change it and make it better so it works for you.

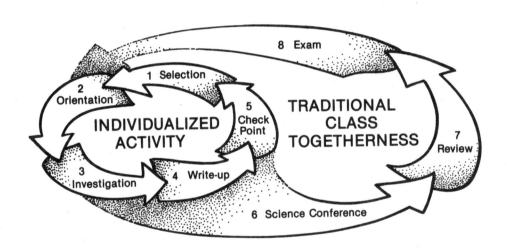

1. SELECTION

Doing TOPS is as easy as selecting the first task card and doing what it says, then the second, then the third, and so on. Working at their own pace, students fall into a natural routine that creates stability and order. They still have questions and problems, to be sure, but students know where they are and where they need to go.

Students generally select task cards in sequence because new concepts build on old ones in a specific order. There are, however, exceptions to this rule: students might *skip* a task that is not challenging; *repeat* a task with doubtful results; *add* a task of their own design to answer original "what would happen if" questions.

2. ORIENTATION

Many students will simply read a task card and immediately understand what to do. Others will require further verbal interpretation. Identify poor readers in your class. When they ask, "What does this mean?" they may be asking in reality, "Will you please read this card aloud?"

With such a diverse range of talent among students, how can you individualize activity and still hope to finish this module as a cohesive group? It's easy. By the time your most advanced students have completed all the task cards, including the enrichment series at the end, your slower students have at least completed the basic core curriculum. This core provides the common background so necessary for meaningful discussion, review and testing on a class basis.

3. INVESTIGATION

Students work through the task cards independently and cooperatively. They follow their own experimental strategies and help each other. You encourage this behavior by helping students only *after* they have tried to help themselves. As a resource person, you work to stay *out* of the center of attention, answering student questions rather than posing teacher questions.

When you need to speak to everyone at once, it is appropriate to interrupt individual task card activity and address the whole class, rather than repeat yourself over and over again. If you plan ahead, you'll find that most interruptions can fit into brief introductory remarks at the beginning of each new period.

4. WRITE-UP

Task cards ask students to explain the "how and why" of things. Write-ups are brief and to the point. Students may accelerate their pace through the task cards by writing these reports out of class.

Students may work alone or in cooperative lab groups. But each one must prepare an original write-up. These must be brought to the teacher for approval as soon as they are completed. Avoid dealing with too many write-ups near the end of the module, by enforcing this simple rule: each write-up must be approved *before* continuing on to the next task card.

5. CHECK POINT

The student and teacher evaluate each write-up together on a pass/no-pass basis. (Thus no time is wasted haggling over grades.) If the student has made reasonable effort consistent with individual ability, the write-up is checked off on a progress chart and included in the student's personal assignment folder or notebook kept on file in class.

Because the student is present when you evaluate, feedback is immediate and effective. A few seconds of this direct student-teacher interaction is surely more effective than 5 minutes worth of margin notes that students may or may not heed. Remember, you don't have to point out every error. Zero in on particulars. If reasonable effort has not been made, direct students to make specific improvements, and see you again for a follow-up check point.

A responsible lab assistant can double the amount of individual attention each student receives. If he or she is mature and respected by your students, have the assistant check the even-numbered write-ups while you check the odd ones. This will balance the work load and insure that all students receive equal treatment.

6. SCIENCE CONFERENCE

After individualized task card activity has ended, this is a time for students to come together, to discuss experimental results, to debate and draw conclusions. Slower students learn about the enrichment activities of faster students. Those who did original investigations, or made unusual discoveries, share this information with their peers, just like scientists at a real conference. This conference is open to films, newspaper articles and community speakers. It is a perfect time to consider the technological and social implications of the topic you are studying.

7. READ AND REVIEW

Does your school have an adopted science textbook? Do parts of your science syllabus still need to be covered? Now is the time to integrate other traditional science resources into your overall program. Your students already share a common background of hands-on lab work. With this shared base of experience, they can now read the text with greater understanding, think and problem-solve more successfully, communicate more effectively.

You might spend just a day on this step or an entire week. Finish with a review of key concepts in preparation for the final exam. Test questions in this module provide an excellent basis for discussion and study.

8. EXAM

Use any combination of the review/test questions, plus questions of your own, to determine how well students have mastered the concepts they've been learning. Those who finish your exam early might begin work on the first activity in the next new TOPS module.

Now that your class has completed a major TOPS learning cycle, it's time to start fresh with a brand new topic. Those who messed up and got behind don't need to stay there. Everyone begins the new topic on an equal footing. This frequent change of pace encourages your students to work hard, to enjoy what they learn, and thereby grow in scientific literacy.

GETTING READY

Here is a checklist of things to think about and preparations to make before your first lesson.

☐ Decide if this TOPS module is the best one to teach next.

TOPS modules are flexible. They can generally be scheduled in any order to meet your own class needs. Some lessons within certain modules, however, do require basic math skills or a knowledge of fundamental laboratory techniques. Review the task cards in this module now if you are not yet familiar with them. Decide whether you should teach any of these other TOPS modules first: *Measuring Length, Graphing, Metric Measure, Weighing* or *Electricity* (before *Magnetism*). It may be that your students already possess these requisite skills or that you can compensate with extra class discussion or special assistance.

☐ Number your task card masters in pencil.

The small number printed in the lower right corner of each task card shows its position within the overall series. If this ordering fits your schedule, copy each number into the blank parentheses directly above it at the top of the card. Be sure to use pencil rather than ink. You may decide to revise, upgrade or rearrange these task cards next time you teach this module. To do this, write your own better ideas on blank 4 x 6 index cards, and renumber them into the task card sequence wherever they fit best. In this manner, your curriculum will adapt and grow as you do.

☐ Copy your task card masters.

You have our permission to reproduce these task cards, for as long as you teach, with only 1 restriction: please limit the distribution of copies you make to the students you personally teach. Encourage other teachers who want to use this module to purchase their *own* copy. This supports TOPS financially, enabling us to continue publishing new TOPS modules for you. For a full list of task card options, please turn to the first task card masters numbered "cards 1-2."

☐ Collect needed materials.

Please see the opposite page.

☐ Organize a way to track completed assignment.

Keep write-ups on file in class. If you lack a vertical file, a box with a brick will serve. File folders or notebooks both make suitable assignment organizers. Students will feel a sense of accomplishment as they see their file folders grow heavy, or their notebooks fill up, with completed assignments. Easy reference and convenient review are assured, since all papers remain in one place.

Ask students to staple a sheet of numbered graph paper to the inside front cover of their file folder or notebook. Use this paper to track each student's progress through the module. Simply initial the corresponding task card number as students turn in each assignment.

☐ Review safety procedures.

Most TOPS experiments are safe even for small children. Certain lessons, however, require heat from a candle flame or Bunsen burner. Others require students to handle sharp objects like scissors, straight pins and razor blades. These task cards should not be attempted by immature students unless they are closely supervised. You might choose instead to turn these experiments into teacher demonstrations.

Unusual hazards are noted in the teaching notes and task cards where appropriate. But the curriculum cannot anticipate irresponsible behavior or negligence. It is ultimately the teacher's responsibility to see that common sense safety rules are followed at all times. Begin with these basic safety rules:

1. Eye Protection: Wear safety goggles when heating liquids or solids to high temperatures.
2. Poisons: Never taste anything unless told to do so.
3. Fire: Keep loose hair or clothing away from flames. Point test tubes which are heating away from your face and your neighbor's.
4. Glass Tubing: Don't force through stoppers. (The teacher should fit glass tubes to stoppers in advance, using a lubricant.)
5. Gas: Light the match first, before turning on the gas.

☐ Communicate your grading expectations.

Whatever your philosophy of grading, your students need to understand the standards you expect and how they will be assessed. Here is a grading scheme that counts individual effort, attitude and overall achievement. We think these 3 components deserve equal weight:

1. Pace (effort): Tally the number of check points you have initialed on the graph paper attached to each student's file folder or science notebook. Low ability students should be able to keep pace with gifted students, since write-ups are evaluated relative to individual performance standards. Students with absences or those who tend to work at a slow pace may (or may not) choose to overcome this disadvantage by assigning themselves more homework out of class.

2. Participation (attitude): This is a subjective grade assigned to reflect each student's attitude and class behavior. Active participators who work to capacity receive high marks. Inactive onlookers, who waste time in class and copy the results of others, receive low marks.

3. Exam (achievement): Task cards point toward generalizations that provide a base for hypothesizing and predicting. A final test over the entire module determines whether students understand relevant theory and can apply it in a predictive way.

Gathering Materials

Listed below is everything you'll need to teach this module. You already have many of these items. The rest are available from your supermarket, drugstore and hardware store. Laboratory supplies may be ordered through a science supply catalog. Hobby stores also carry basic science equipment.

Keep this classification key in mind as you review what's needed:

special in-a-box materials:	general on-the-shelf materials:
Italic type suggests that these materials are unusual. Keep these specialty items in a separate box. After you finish teaching this module, label the box for storage and put it away, ready to use again the next time you teach this module.	Normal type suggests that these materials are common. Keep these basics on shelves or in drawers that are readily accessible to your students. The next TOPS module you teach will likely utilize many of these same materials.
(substituted materials):	*optional materials:
A parentheses following any item suggests a ready substitute. These alternatives may work just as well as the original, perhaps better. Don't be afraid to improvise, to make do with what you have.	An asterisk sets these items apart. They are nice to have, but you can easily live without them. They are probably not worth the an extra trip, unless you are gathering other materials as well.

Everything is listed in order of first use. Start gathering at the top of this list and work down. Ask students to bring recycled items from home. The teaching notes may occasionally suggest additional student activity under the heading "Extensions." Materials for these optional experiments are listed neither here nor in the teaching notes. Read the extension itself to find out what new materials, if any, are required.

Needed quantities depend on how many students you have, how you organize them into activity groups, and how you teach. Decide which of these 3 estimates best applies to you, then adjust quantities up or down as necessary:

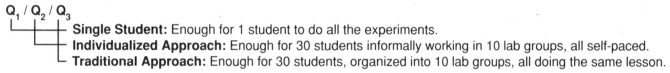

Q_1 / Q_2 / Q_3
- **Single Student:** Enough for 1 student to do all the experiments.
- **Individualized Approach:** Enough for 30 students informally working in 10 lab groups, all self-paced.
- **Traditional Approach:** Enough for 30 students, organized into 10 lab groups, all doing the same lesson.

KEY:	*special in-a-box materials* (substituted materials)	general on-the-shelf materials *optional materials

Q_1 / Q_2 / Q_3		Q_1 / Q_2 / Q_3	
10/300/300	sheets of lined notebook paper	1/10/10	margarine lids (panes of glass)
1/15/15	rulers (straight edges)	1/3/10	*sea shells (egg shells)*
.1/1/1	*quart of coarse brown sand with flecks of mica*	1/2/5	needle-nose pliers
1/8/10	*spoons	1/30/30	a large coffee can or equivalent
1/8/10	pie tins	1/30/30	safety goggles (twice this many plastic produce bags)
1/30/30	hand lenses	1/1/1	a place where students can search for rocks
1/1/1	box of granulated sugar	1/1/1	*a commercial collection of common rocks*
1/10/10	microscope slides	1/5/10	scissors
1/10/10	a candle with drip catcher and matches (Bunsen burners)	1/30/30	egg cartons
2/20/20	*pieces (each) of granite and basalt, about golf ball size*	.5/2/5	cups of oil-based clay —1 cup = 1/2 lb
3/10/12	glass jars	1/10/10	gram balances
1/10/10	eyedroppers	3/30/30	paper towels
1/1/1	package of table salt	1/10/10	100 ml graduated cylinders
1/15/30	pennies	1/1/1	box of pepper
2/15/20	paper clips	1/10/10	small beakers or jars
1/10/10	*common bricks*	1/10/10	*hand calculators
2/20/20	pieces of chalk	5/15/50	glass marbles
1/1/1	roll of masking tape	1/4/10	*empty film canisters with snap on lids*
2/10/20	test tubes	1/1/5	large wash tubs (buckets)
1/10/10	dropper bottles with 5% hydrochloric acid	1/1/1	a freezer (freezing weather)
		.5 /5/5	*cups Epsom salts*
		1/3/10	stirring rods
		1/2/10	dictionaries

Sequencing Task Cards

    This logic tree shows how all the task cards in this module tie together. In general, students begin at the trunk of the tree and work up through the related branches. As the diagram suggests, the way to upper level activities leads up from lower level activities.

    At the teacher's discretion, certain activities can be omitted or sequences changed to meet specific class needs. The only activities that must be completed in sequence are indicated by leaves that open *vertically* into the ones above them. In these cases the lower activity is a prerequisite to the upper.

    When possible, students should complete the task cards in the same sequence as numbered. If time is short, however, or certain students need to catch up, you can use the logic tree to identify concept-related *horizontal* activities. Some of these might be omitted since they serve only to reinforce learned concepts rather than introduce new ones.

    On the other hand, if students complete all the activities at a certain horizontal concept level, then experience difficulty at the next higher level, you might go back down the logic tree to have students repeat specific key activities for greater reinforcement.

    For whatever reason, when you wish to make sequence changes, you'll find this logic tree a valuable reference. Parentheses in the upper right corner of each task card allow you total flexibility. They are left blank so you can pencil in sequence numbers of your own choosing.

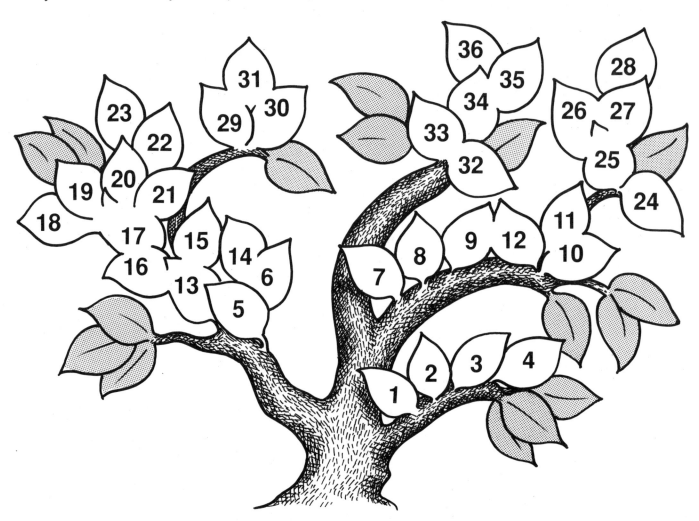

ROCKS & MINERALS 23

E

LONG-RANGE OBJECTIVES

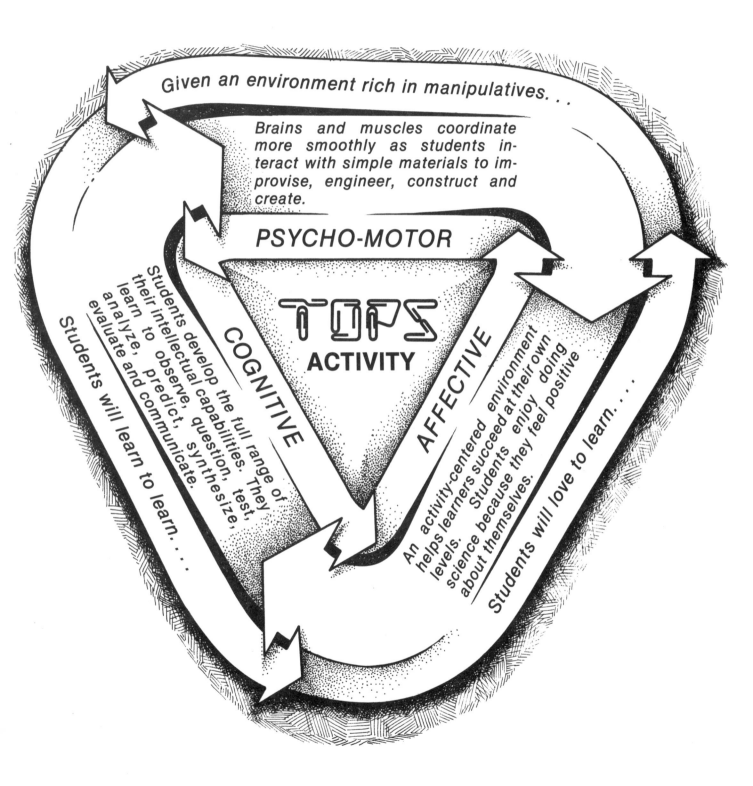

Given an environment rich in manipulatives. . .

Brains and muscles coordinate more smoothly as students interact with simple materials to improvise, engineer, construct and create.

PSYCHO-MOTOR

TOPS ACTIVITY

COGNITIVE

Students develop the full range of their intellectual capabilities. They learn to observe, question, test, analyze, predict, synthesize, evaluate and communicate.

Students will learn to learn. . . .

AFFECTIVE

An activity-centered environment helps learners succeed at their own levels. Students enjoy doing science because they feel positive about themselves.

Students will love to learn. . . .

Review / Test Questions

task 1
This bar graph shows the relative abundance of elements in the earth's crust.

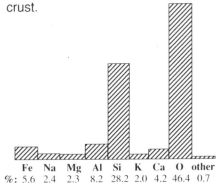

	Fe	Na	Mg	Al	Si	K	Ca	O	other
%:	5.6	2.4	2.3	8.2	28.2	2.0	4.2	46.4	0.7

List the first 8 elements by name. Start with the most common element at the top of your list and work down in order of abundance.

task 2
Write "feldspar", "mica" or "quartz" after each descriptive word that applies. A word may have 1, 2 or 3 minerals written after it.

translucent:
shiny:
opaque: cleaves:
fractures:
found in sand:
flat:
chunky:
streaks across brick:
scratches brick:

task 3
Transparent quartz grains are gathered from sand. Half are poured into a jar labeled "quartz grains." The other half are smashed with a hammer and poured into a jar labeled "quartz powder."

Describe how the crystals in each jar look. Explain why you think so.

task 4
One of these rocks is volcanic and one is plutonic. Explain how you know which is which.

A: B:

task 5
Place minerals X, Y, and Z on a scale of hardness:
  X scratches **Y** and **Z**
  Y streaks across **Z**

task 6
Which is harder, the lead in your pencil or the paper you write on? Explain.

task 7
You find 2 rocks that both broke off the same parent slab of granite. Rock (A) is rough and jagged. Rock (B) is round and smooth. Propose a theory that explains these differences.

A: B:

task 8
One of these core samples was taken in the mountains, the other in a broad river valley. Which sample most likely came from where?

A: 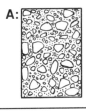 B: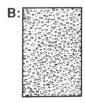

task 9
Explain how clastic sedimentary rock forms. Give an example.

task 10

A:

B:

Which mountain is more likely made from limestone? from granite? Explain.

task 11

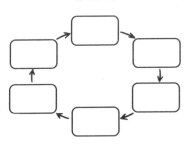

Each phrase below describes a step in the calcium carbonate cycle. Write a letter in each box so the phrases are arranged in the correct order.

A. absorbed by marine animals
B. uplifted to limestone mountains
C. washed into the sea
D. buried deeply, layer upon layer
E. shelled remains deposited on sea bed
F. chemical erosion by rain

task 12
Compare the cement that holds sandstone together with the cement that binds limestone. How can you identify each?

task 13
Why is it helpful to break rocks open when you study them? What safety precaution should you observe?

task 14
Streak testing a rock may be more complicated than streak testing a mineral. Why is this so?

task 15
Draw a massive texture on rock A; a foliated texture on rock B. Explain the difference.

A:

B:

G

Answers

Photocopy all questions to the left. On a separate sheet of blank paper, cut and paste all those boxes you want to use as test questions. Include questions of your own design, as well. Crowd all these questions onto a single page for students to answer on their own papers, or leave space after each question for student responses, as you wish. Duplicate a class set and your custom-made test is ready to use. Gather all leftover questions to use as a review in preparation for the final exam.

task 1
1. oxygen
2. silicon
3. aluminum
4. iron
5. calcium
6. sodium
7. magnesium
8. potassium

task 2
translucent: quartz, (mica)
shiny: mica
opaque: feldspar, quartz
cleaves: mica, (feldspar)
fractures: feldspar, quartz
found in sand: feldspar, quartz, mica
flat: mica
chunky: feldspar, quartz
streaks across brick: mica
scratches brick: feldspar, quartz

task 3
The quartz grains are clear and sparkling (similar to granulated sugar) because the crystals are large enough to both transmit and reflect light. The quartz powder is white and opaque (similar to powdered sugar) because the crystals are too small and rough to transmit or reflect light.

task 4
Rock **A** has large crystals. It is likely plutonic since its crystals had time to cool slowly an underground pluton and grow relatively large. Rock **B** has small crystals. It is likely volcanic since its crystals cooled too rapidly, in a surface eruption, to grow very large.

task 5

←soft—Y—Z—X—hard→

task 6
Pencil lead is softer than paper, because it leaves a streak of graphite when you rub it across paper.

task 7
Rock **B** was rounded and smoothed by heavy erosion due to wind and water. Rock **A** escaped this erosion, possibly by breaking off the parent rock at a much later time; or perhaps it was well protected from the elements through light burial.

task 8
Core **A** likely came from the mountains. The clasts are still unsorted by size with large and small rocks all deposited in a jumble. Core **B** likely came from the river valley, because the sediment has been well sorted into uniform small clasts that were small enough to have been transported great distances and deposited together, perhaps by an ancient river.

task 9
Rocks gradually erode mechanically, breaking into smaller and smaller pieces. These clasts are transported by wind and water over great distances and finally deposited into beds of sediment. Here minerals that are dissolved in water gradually fill up the tiny spaces between the clasts, eventually cementing them into rock. Granite, for example, erodes into sand particles that cement into sandstone.

task 10
Mountain **A** is probably made from limestone because this mineral is easily smoothed and channeled by chemical weathering. Mountain **B** is likely made from granite because this mineral is more resistant to chemical erosion, displaying the jagged edges and clastic debris characteristic of mechanical weathering.

task 11
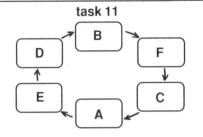

task 12
Individual clasts of sand are usually cemented by silica (SiO_2). It is characteristically hard and tough, and won't fizz in dilute HCl. Limestone is cemented by calcium carbonate ($CaCO_3$). It is softer than silica and bubbles in dilute HCl.

task 13
The inside texture of a rock is unaltered by chemical and mechanical erosion, allowing easier mineral recognition. When breaking open rocks, always capture potential flying fragments in a bag, or wear safety goggles, to reduce the risk of eye injury.

task 14
A mineral has only 1 specific hardness, but a rock may contain many different minerals, each with a different hardness. Cements that bind these minerals together may be hard or soft, as well.

task 15

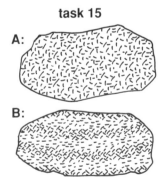

A:

B:

Massive rocks have no particular pattern: their textures are random and nonrepeating. Foliated rocks display some kind of a pattern; their textures are both ordered and repeating.

Review / Test Questions (continued)

task 16-21
Write the name of each rock in the correct egg-cup category below:

chalk, gabbro, ore bearing, pumice, serpentinite, gneiss, shale, flint, gypsum, sandstone, quartzite, andesite.

1 CLASTIC SEDIMENTARY coarse grained	**7** PLUTONIC IGNEOUS coarse crystals
2 CLASTIC SEDIMENTARY fine grained	**8** VOLCANIC IGNEOUS fine crystals
3 CLASTIC SEDIMENTARY poorly sorted	**9** VOLCANIC IGNEOUS pyroclastic
4 CHEMICAL SEDIMENTARY carbonate	**10** MASSIVE METAMORPHIC
5 CHEMICAL SEDIMENTARY silica	**11** FOLIATED METAMORPHIC
6 CHEMICAL SEDIMENTARY other	**12** OTHER METAMORPHIC

task 18-21
Name each rock. Then write a word or phrase to describe how it forms:

A. A dark-colored, fine grained rock with gas holes
B. A light-colored rock with coarse interlocking flecks of quartz, feldspar and mica
C. A hard, black rock with a glassy texture
D. A soft, black rock made from carbon
E. A soft rock containing sea shells
F. A smooth, banded, translucent rock that is very hard
G. A rock that bubbles in dilute HCl and has coarse interlocking crystals
H. A light-colored rock foliated with thin layers of mica
I. A clastic rock with small, well-sorted clasts
J. A clastic rock with poorly-sorted clasts

task 22
A. Fill each box in this rock cycle with the correct rock type: metamorphic rock, sedimentary rock, molten magma or igneous rock.

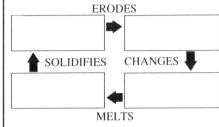

B. Which steps of this cycle can move in both directions? Explain.

task 23
Name the…
A. 2 most common igneous rocks.
B. 3 most common sedimentary rocks.
C. 3 most common metamorphic rocks.

task 24
You are given 2 different rocks. Tell how you would use a gram balance to find which rock is most permeable to water.

task 25
A nugget of pure gold raises the water in a graduated cylinder from 50 ml to 57 ml. It has a mass of 135.1g.
A. Find the volume of this nugget. Show your math.
B. Find the density of gold. Show your math.

task 26
Answer the previous question first, before you try this one.
A. 10 ml of liquid mercury has a mass of 136 g. Find the density of mercury.
B. Liquid mercury is poured over gold-bearing granite that has been thoroughly crushed. Explain how this separates out the gold.

task 27
Taffy is much denser than cotton candy, yet they are both made from melted sugar. Explain why these 2 forms of sugar have different densities.

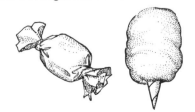

task 28
Ice has a density of .92 g/ml. What is its specific gravity?

task 29
A jar is completely filled with water and sealed with a screw-on lid.

A. Predict what happens to this jar if you leave it in a freezer overnight.
B. How does this experiment model mechanical erosion in nature?

task 30
Salt crystals can be made by dissolving table salt in hot water, then evaporating away the water. List 2 things you can do to grow the largest crystals possible.

task 31
You discover a rich vein of pegmatite running through granite rock. Explain how it got there.

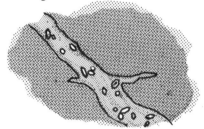

task 32
On an archaeological dig, you unearth a log that has been turned to stone. How did this happen?

task 33
How is a stalactite similar to an icicle? How is it different?

task 34
Silica ions (+4) and oxygen ions (-2) combine into basic building blocks called tetrahedra. Describe a single tetrahedron in words and pictures. Compute its net charge.

task 35
Silica tetrahedra combine to form both flat sheets and 3-dimensional frameworks. Which way do these tetrahedra join in mica? in quartz? Explain your reasoning.

task 36
Most rock forming minerals contain silica tetrahedra. List several that contain NO silica tetrahedra.

Answers (continued)

task 16-21

1 sandstone	**7** gabbro
2 shale	**8** andesite
3 ore-bearing	**9** pumice
4 chalk	**10** quartzite
5 flint	**11** gneiss
6 gypsum	**12** serpentinite

task 18-21

A. basalt: igneous, volcanic
B. granite: igneous, plutonic
C. obsidian: igneous, pyroclastic
D. coal: sedimentary, deposited plant remains
E limestone: sedimentary, calcium carbonate deposited as seashells
F. agate: sedimentary, precipitated silica
G. marble: metamorphic, comes from limestone
H. schist: metamorphic, comes from mica-rich clay, sand or granite
I. sandstone: sedimentary, clasts transported long distances before deposition
J. conglomerate: sedimentary, clasts consolidated early

task 22

A.

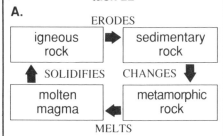

B. Magma that *solidifies* as igneous rock can remelt to magma. Sedimentary rock that *changes* to metamorphic rock can erode back to sedimentary rock.

task 23

A. basalt, granite
B. shale, limestone, sandstone
C. gneiss, schist, marble

task 24

First weigh both rocks dry. Then soak them in water and reweigh both rocks wet. The rock that gains the most mass has the greatest permeability. (More accurately, the rock that gains the greatest *percentage* of mass has the greatest permeability.)

task 25

A. volume = 57 ml - 50 ml = 7 ml
B. density = 135.1 g / 7 ml = 19.3 g/ml

task 26

A. density = 136 g / 10 ml = 13.6 g/ml
B. Mercury is less dense than gold, more dense than granite. Thus all the granite would float to the top of the mercury, leaving pure gold at the bottom.

task 27

The same mass of melted sugar occupies a much greater volume when spun into thin glassy threads of cotton candy than when cooled into solid squares of taffy. All the extra air space gives cotton candy much less density.

task 28

s.g. = .92 g/ml ÷ 1.0 g/ml = .92

task 29

A. The water would freeze and expand inside the jar, cracking open the glass.
B. In a similar manner, water penetrates between cracks in rocks, then freezes and expands to break the rock apart.

task 30

1. Allow the crystals plenty of time to grow large through slow cooling and slow evaporation.
2. Dissolve as much salt as possible in the water (saturate the solution).

task 31

Granite plutons cool and solidify from the outside in, concentrating highly volatile minerals (those that solidify last) into the still-liquid magma center. As cracks form in the walls of the contracting pluton, mineral-rich magma flows in from this center to fill them up, forming rich veins of pegmatite.

task 32

Rainwater chemically eroded away rock, transporting dissolved silica long distances. Some of this water soaked into the old log, dissolving away its wood fiber and replacing it with silica. Over a long period of time, the log petrified to hard silica.

task 33

A stalactite grows down from above, accumulating new calcium carbonate on its growing tip just as an icicle grows from the top down, adding new ice to its growing tip. The basic difference is that calcium carbonate precipitates from solution as its water evaporates away, leaving a permanent solid that will last for centuries. Water on an icicle simply freezes. It will warm and melt away with tomorrow's sun.

task 34

Each silicon ion is surrounded by 4 oxygen ions to form a 4-sided pyramid that looks like this.

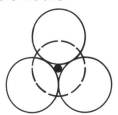

The chemical formula for this unit is SiO_4. Its net charge is:

$$1(+4) + 4(-2) = -4$$

task 35

Mica is flat, thin and flaky so the tetrahedra must join in flat sheets. Quartz by contrast is solid, chunky and hard, suggesting a 3-dimensional framework structure.

task 36

Pure forms of limestone ($CaCO_3$), coal (C), rock salt (NaCl) and gypsum ($CaSO_4 \cdot H_2O$) contain no silica.

TEACHING NOTES
For Activities 1-36

Task Objective (TO) appreciate that the earth's crust is made almost entirely from 8 basic elements. To graph these elements on a bar graph.

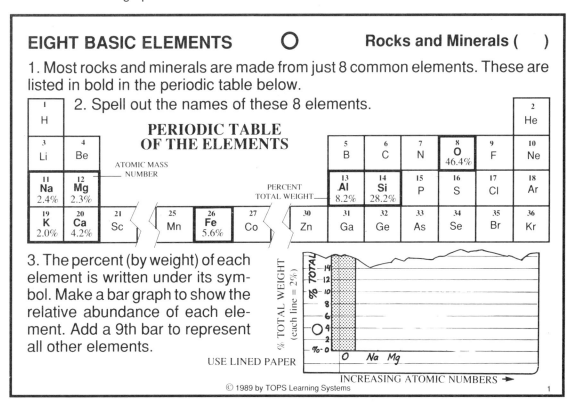

EIGHT BASIC ELEMENTS ○ Rocks and Minerals ( )

1. Most rocks and minerals are made from just 8 common elements. These are listed in bold in the periodic table below.

2. Spell out the names of these 8 elements.

PERIODIC TABLE OF THE ELEMENTS

3. The percent (by weight) of each element is written under its symbol. Make a bar graph to show the relative abundance of each element. Add a 9th bar to represent all other elements.

USE LINED PAPER

INCREASING ATOMIC NUMBERS →

© 1989 by TOPS Learning Systems   1

Answers / Notes

2. *Students may need to consult a physics or chemistry text to correctly list all the names. Or they can look up abbreviations in the back of a dictionary.*

oxygen	silicon
sodium	potassium
magnesium	calcium
aluminum	iron

3. *These elements are graphed in order of atomic mass numbers. This ordering (from left to right and from top to bottom) is natural to anyone that reads. Other orderings are OK too.*

Scale is always an important consideration when graphing. We have chosen 1 line = 2%. Other scales are equally possible. Students who use a scale of 1 line = 1% must tape 2 pieces of notebook paper together to accomodate the long oxygen bar. A scale of 1 line = 5%, by contrast, fits on less than a half sheet of paper.

Your students will probably ask you a whole host of questions: What order? What scale? How wide should the bar be? Allow them this grand opportunity for creative problem solving. Just smile and say, "you decide."

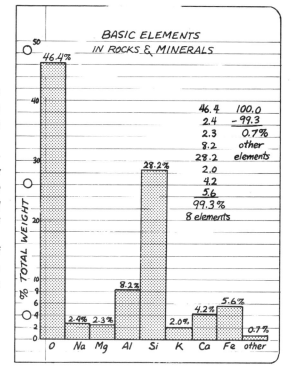

Materials

☐ Lined notebook paper.
☐ A ruler or straight edge.

(TO) sort grains of sand by shape, opacity and color. To identify 3 common minerals in sand.

TINY TREASURES ◯ Rocks and Minerals (   )

1. Put about a spoonful of sand into a pie tin. Use a hand lens to inspect the tiny grains. Pick out different shapes, lusters and colors with the moistened tip of your pencil.

2. Copy this mineral table. For each box, find the correct grain sample, describe it, then tape it in.

CHUNKY TRANSLUCENT Quartz	description: *crystal clear* specimen: ◯		
CHUNKY OPAQUE Quartz or Feldspar			
FLAT SHINY Mica			

3. Minerals *cleave* (break evenly) to form flat, smooth surfaces. Describe sand grains you see that have cleaved.

4. Minerals *fracture* (break unevenly) to form irregular, rough surfaces. Describe sand grains you see that have fractured.

© 1989 by TOPS Learning Systems   2

Introduction

Consider how different minerals have different lusters. Distinguish between a *translucent* mineral that allows the passage of light, an *opaque* mineral that blocks the passage of light, and a *shiny* mineral that reflects light.

Answers / Notes

2. Descriptions and specimens will vary widely.

CHUNKY TRANSLUCENT Quartz	*crystal clear*	*milky yellow*	*clear brown*
CHUNKY OPAQUE Quartz or Feldspar	*bone white*	*pink*	*black*
FLAT SHINY Mica	*white, translucent*	*brown translucent*	*black*

3. Mica cleaves into smooth flat mineral grains.

4. Quartz and most feldspar fracture into chunky irregular pieces. Some grains of feldspar may show a smooth cleavage if the surface has not been excessively eroded away.

Materials

☐ Coarse brown sand, derived from granite. This will contain the required quartz, opaque feldspar and shiny mica. Fine beach sand should not be substituted except as a last resort. Its heavily eroded grains are too small for really good observation with a hand lens, and all traces of mica and feldspar have probably long since eroded away. Black or dark grey sand decomposed from lava and other basalt is equally unsuitable. If you live in an area that lacks granite, ask a far-away friend to mail you a quart.

☐ A spoon (optional).
☐ A pie tin.
☐ A hand lens.

(TO) recognize that one mineral may assume different solid forms. To account for the great diversity of rocks in the earth's crust.

ONE MINERAL / MANY FORMS ○   Rocks and Minerals (   )

1. Divide a small pinch of sugar into 3 tiny piles on a glass slide.

LARGE CRYSTALS
Leave this pile alone.

SMALL CRYSTALS
Grind to a fine powder with the back of your finger nail.

FUSED MASS
Heat gently until the sugar *just* melts.

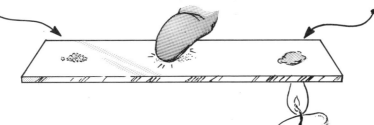

2. Examine each pile of sugar with a hand lens.
   a. Describe what you see.
   b. Propose a theory to explain why each pile of sugar looks the way it does.

3. A huge variety of different-looking rocks contain the same basic mix of just 8 elements. How does this experiment help explain such diversity?

3

Answers / Notes

1. *Students should heat the sugar very gently and remove it from the candle flame as soon as it melts. Otherwise it will quickly blacken, oxidizing into elementary carbon.*

2. *Remember that part (b) asks for each student's best guess. As such, let all reasonable answers stand.*

LARGE CRYSTALS	SMALL CRYSTALS	FUSED MASS
a. The sugar is colorless and translucent with a sparkling luster. b. Light travels through the large whole crystals and also reflects off of flat smooth crystal faces.	a. The sugar is white and opaque with a dull luster. b. The crystals are too small and rough to transmit or reflect light.	a. The sugar is a clear, smooth, sticky, transparent solid. b. The crystals melted into a liquid that did not recrystalize.

3. If one mineral, sugar, can look so radically different all by itself, it is easy to understand by analogy how 8 elements in differing combinations and amounts, and subjected to different kinds of external forces, can produce so much more diversity.

Materials
☐ Granulated sugar.
☐ A glass slide.
☐ A candle with drip catcher and matches.

(TO) observe that coarse crystals form more slowly than fine crystals. To deduce the origin of granite and basalt based on the size of their crystals.

IGNEOUS ROCK ○ Rocks and Minerals ( )

1. Lay down a row of saturated salt water drops across the diameter of an aluminum pie tin. Rest it on a jar so a candle fits under the overhang, directly below the first drop.

    a. Light the candle. Its flame should touch the bottom edge of the pie tin.
    b. Write your observations over the next 10 to 20 minutes.
    c. How is the size of a crystal related to how fast it forms?

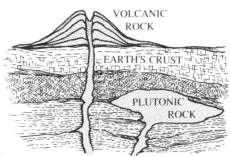

2. Examine 2 samples of igneous rock — granite and basalt. Use a hand lens.

    a. Compare and contrast the crystallized minerals in each rock.
    b. Which rock is most likely *plutonic*? Most likely *volcanic*? Explain how you know.

4

Introduction

Review the 2 basic ways that igneous rock crystallizes from molten magma: If it cools slowly over thousands of years in deep underground bodies (plutons or batholiths), it is called *plutonic*. If it erupts through cracks in the earth's crust to cool rapidly at the surface, it is called *volcanic*.

Show your class samples of granite and basalt rock, but don't say which is plutonic and which is volcanic. Students will decide this in step 2.

Answers / Notes

1b. The water drops evaporate one by one, leaving salt crystals behind. Near the flame this happens so rapidly, the crystals sizzle and splatter. Farther from the flame, the drops evaporate more and more slowly.

1c. The crystals that formed rapidly are too small to see individually. (Their dull white appearance is similar to crushed sugar.) The crystals that formed more slowly are much larger. (Their uniform shape and translucence is similar to granulated sugar.)

2a. Granite has lightly colored flecks of crystallized minerals. Translucent quartz, opaque feldspar and black mica predominate. The crystals in basalt are much finer and more darkly colored, but still sparkle in strong light.

2b. Granite must be plutonic: its crystal flecks are large and well-formed, indicating it cooled slowly in an underground pluton or batholith. Basalt must be volcanic: its finer crystal ground mass indicates it must have erupted to the surface and cooled rapidly.

Materials

☐ Freshly broken granite and basalt rocks, about golf-ball size. Both can be ordered from scientific supply companies, but since they are the most abundant rocks on earth, you may have them in your own back yard or gravel driveway. If not, keep these rocks in mind when you travel, or ask a friend to mail you some. Five pounds of each specimen, broken into smaller pieces, should accommodate an entire class.

☐ A hand lens.

☐ A candle with drip catcher and matches.

☐ A glass jar, pie tin and eyedropper.

☐ A saturated solution of salt water. Make this in advance. Stir excess salt into a quart of water. Allow it to settle overnight. Then pour off the clear solution into another container. Seal and label.

(TO) interpret the hardness or softness of objects by scratching one against the other. To construct a scale of relative hardness for common objects.

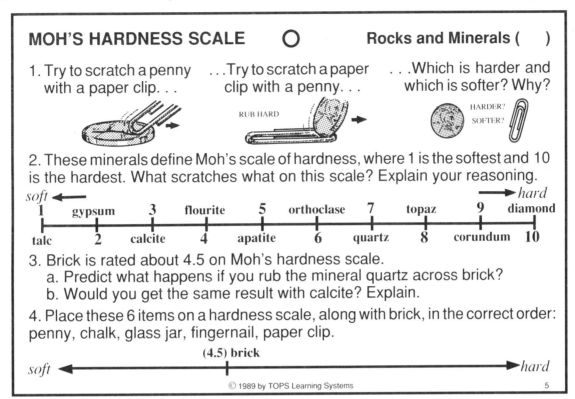

1. Try to scratch a penny with a paper clip. . .   . . .Try to scratch a paper clip with a penny. . .   . . .Which is harder and which is softer? Why?

RUB HARD

HARDER?
SOFTER?

2. These minerals define Moh's scale of hardness, where 1 is the softest and 10 is the hardest. What scratches what on this scale? Explain your reasoning.

soft ← → *hard*

1 — gypsum — 3 — flourite — 5 — orthoclase — 7 — topaz — 9 — diamond

talc — 2 — calcite — 4 — apatite — 6 — quartz — 8 — corundum — 10

3. Brick is rated about 4.5 on Moh's hardness scale.
   a. Predict what happens if you rub the mineral quartz across brick?
   b. Would you get the same result with calcite? Explain.

4. Place these 6 items on a hardness scale, along with brick, in the correct order: penny, chalk, glass jar, fingernail, paper clip.

(4.5) brick

soft ← → *hard*

© 1989 by TOPS Learning Systems

5

Answers / Notes

1. The harder paper clip scratches the softer penny.

2. Harder objects scratch softer objects. Thus a mineral will scratch anything to its left and be scratched by anything to its right.

3a. Quartz at 7 is harder than brick at 4.5. Thus quartz scratches brick.

3b. No. Calcite at 3 is softer than brick at 4.5. Thus brick scratches calcite.

4.

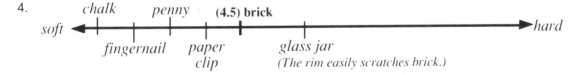

chalk   *penny*   **(4.5) brick**

soft ← → *hard*

fingernail   *paper clip*   *glass jar*
(The rim easily scratches brick.)

Materials

☐ Test materials: a penny, paper clip, brick, chalk and glass jar.

(TO) interpret the hardness or softness of objects by streaking them across brick.

STREAK TEST Rocks and Minerals ( )

1. Draw chalk across the surface of a brick. How does this "streak test" reveal the hardness of chalk relative to brick?

2. Streak a penny and a glass jar across the brick. Write your observations and conclusions.

3. Double over a strip of masking tape about 4 times.

4. Use it to pad your finger while you streak different grains of sand across brick. Then fill in the table.

	Scratch or streak?	Conclusion?
QUARTZ		
FELDSPAR		
MICA		

6

Answers / Notes

1. Chalk leaves a streak (a powdered trail) across brick. This shows that it is softer than brick.

2. The penny leaves a copper-colored streak across brick, whereas the glass jar scratches the brick. Thus the penny is softer than brick and the glass jar is harder than brick.

4.

	Scratch or streak?	Conclusion?
QUARTZ	scratch	harder than brick
FELDSPAR	scratch	harder than brick
MICA	streak	softer than brick

Materials
☐ Test materials: a brick, chalk, penny, glass jar.
☐ Masking tape.
☐ Coarse sand with flecks of mica.

(TO) mechanically erode pieces of granite by rubbing them together. To understand how this happens in nature, changing the appearance of rock and producing soil.

MECHANICAL WEATHERING   Rocks and Minerals (   )

Grind 2 granite rocks together as hard as you can for about 1 minute. Collect the falling fragments on clean notebook paper, then tap them together into one central pile.

1. Describe the tiny fragments that eroded from the parent rock. Can you still recognize them as crystals?

2. How has the texture of the parent rock changed? Propose a theory to explain why rocks look smooth and dull on the outside.

3. How do rocks mechanically weather in nature? List as many natural erosion forces as you can.

4. Where does topsoil come from? Why is it important to conserve topsoil?

© 1989 by TOPS Learning Systems   7

Answers / Notes

1. The eroded rock looks like gray powder. Only an occasional larger fragment contains crystals that can be recognized.

2. The surface of each parent rock takes on a duller, smoother, more uniform light-grey appearance at the point of contact, hiding the original crystal texture underneath. As rocks erode in nature through the action of wind and water, surface features are similarly smoothed over and greyed-out.

3. • water freezing and expanding inside rock cracks and crevices
    • sand blasting by wind
    • growing roots
    • burrowing animals
    • lightning
    • earthquakes
    • rock slides
    • moving water: floods, rivers, tides
    • human activities

4. Topsoil comes from eroding rock. It should be conserved because it forms so slowly. *(Rock erosion is estimated, on average, to produce 1 inch of topsoil every 9,000 years.)*

Materials

☐ Two granite rocks. They should be large enough to grasp easily and rub together.

(TO) study how wind and water sort rock clasts by size and deposit them into graded beds.

TRANSPORT AND SORTING ○   Rocks and Minerals ( )

1. Grind 2 granite rocks together for about 1 minute, as before. Collect the *clasts* (rock fragments) on clean notebook paper.

2. Hold the paper like a funnel and pour the rock fragments into a test tube. Notice how gravity sorts these clasts by size as they move down the paper.
     a. Write your observations.
     b. Explain how wind has a similar sorting effect.

CLASTS SORTED BY GRAVITY

3. Add a few drops of water to the test tube and shake it up.
     a. Write your observations.
     b. Discuss the role that rivers play in transporting and sorting sediment.

4. Add a centimeter of sand to 2 test tubes. Fill one with water; leave the other filled with air. Then vigorously shake both test tubes.
     a. After 1 minute, make labeled drawings of each one.
     b. Contrast the *sorted* sediment with *unsorted* sediment.

WATER   AIR

1 cm

© 1989 by TOPS Learning Systems   8

Answers / Notes

2a. The larger clasts roll down the paper most easily, leaving finer grains and dust behind.
2b. Wind has a similar but opposite sorting effect. It most easily carries away the smaller, lighter clasts, leaving the coarser grains behind.

3a. The water turns cloudy with a suspension of very fine sediment.
3b. Fine sediments that mix with river water are transported over great distances. Coarse sediments that remain at the river bottom move much more slowly. Larger heavier rocks move hardly at all. Hence, over time, rock fragments of the same size are deposited together — larger clasts upstream, finer clasts downstream.

4a.

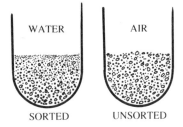

WATER   AIR

SORTED   UNSORTED

4b. The test tube with water has sorted the sand into a graded bed: the largest clasts are at the bottom, working up to a top layer of very fine clasts. The test tube with air, by contrast, has an unsorted bed: all clast sizes remain randomly mixed throughout.

Materials

☐ Two granite rubbing rocks.
☐ A test tube.
☐ Coarse sand. Fine-grained beach sand is not a good substitute, unless you pulverize part of it to provide a wider range of grain sizes.

(TO) mechanically generate a new sedimentary rock from preexisting igneous rock.

SEDIMENTARY ROCK ○ Rocks and Minerals ( )

1. Grind 2 granite rocks together for about 1 minute as before. Collect the clastic debris on clean notebook paper.

2. Transfer the sediment to one end of a microscope slide. Add just *one* drop of dilute hydrochloric acid and mix with a paper clip to form a tiny mud puddle.

3. Gently heat to dryness over a candle flame, then allow the slide to cool.

4. Try to blow the dried mud off the slide. Notice that you have begun to form a new rock.
  a. Why is "clastic sedimentary rock" a good name for this new material?
  b. Compare the texture of this new rock with its parent rock.

5. Dissolved minerals in the acid drop *cement* the clasts in your new rock together. Brush the slide lightly with your fingertips to discover where this cementation is the strongest.

9

Answers / Notes

4a. The new material is made from *clasts* of preexisting rock that accumulate as *sediment* and harden as *rock*. The result is logically called "clastic sedimentary rock."

4b. The new rock has a somewhat uniform, dull, gray texture. Its mineral fragments are generally too small and broken to sparkle with much reflected light or to show distinct colors. The parent rock made from igneous granite, by contrast, has a coarse interlocking crystal texture. Its mineral flecks are large enough to reflect light and show distinct color.

5. A distinct ring shape is left on the slide after lightly brushing the sediment. This suggests that the new rock is most strongly cemented around its outside edge. *(This is where the acid drop evaporates last, depositing the strongest concentration of cementing minerals.)*

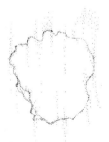

Materials

☐ Two granite rubbing rocks.
☐ A microscope slide.
☐ A 5% solution of hydrochloric acid in labeled dropper bottles. Dilute 10 ml concentrated HCl with 190 ml water, then pour the dilute acid into labeled dropper bottles for student use. If HCl is unavailable you may substitute other strong acids diluted to the same safe concentration. Phosphoric acid, for example, is generally sold in the aquarium section of pet stores.
☐ A paper clip.
☐ A candle with drip catcher and matches.

(TO) dissolve a piece of chalk in weak acid and observe the resulting erosion patterns. To contrast chemical weathering with mechanical weathering.

CHEMICAL WEATHERING   Rocks and Minerals ( )

1. Stand a short piece of blackboard chalk on a flat lid.

2. Surround it with about 4 eye-droppers full of 5% hydrochloric acid.

3. Calcium carbonate (chalk) dissolves in acid as it produces carbon dioxide gas. What evidence for this reaction do you *see* and *hear*?

4. Examine the sides of the chalk above the water line with a hand lens.
   a. Draw the tiny drainage patterns that open up as the calcium carbonate dissolves away.
   b. Whole mountains are made from calcium carbonate (limestone). Draw a picture to show chemical weathering on this grander scale.

5. How is this chemical weathering of chalk different from the mechanical weathering of granite you studied earlier?

© 1989 by TOPS Learning Systems   10

Answers / Notes

3. Gas bubbles arise on the surface of the chalk where it comes in contact with the weak acid. There is a distinct fizzing sound as gas bubbles continuously pop and new ones form.

4.   a.    b.

5. Weak acid dissolved the chalk chemically and carried it away in solution. It was not mechanically broken into smaller pieces like granite that was rubbed together in previous experiments.

Materials

☐ Blackboard chalk.

☐ Any flat surface made from glass, plastic or painted metal. Coated canning jar lids or panes of glass work well. Watch glasses have too much curve to work properly. Unpainted metal surfaces are unsuitable because they corrode in weak acid.

☐ A dropper bottle of 5% HCl.

(TO) trace the path of calcium carbonate as it travels from limestone to water, to living organisms, to shelled remains and finally recycles back to chalk and limestone.

SEA FLOOR SEDIMENT ⚪ Rocks and Minerals (   )

1. Examine the chalk that was weathered by 5% hydrochloric acid.

    a. Draw and label a "before" and "after" picture.

    b. Account for the changes in each picture.

BEFORE   AFTER

2. Put a drop of 5% hydrochloric acid on a sea shell.

    a. Write your observations.

    b. What are shells made from? Where do you think sea animals might get this compound to make their shells?

    c. What happens to their shelled remains after these animals die?

3. Blackboard chalk is an accumulation of the shells of microscopic sea animals. Show by diagram how the chalk in your classroom became a chemical sedimentary rock.

© 1989 by TOPS Learning Systems   11

Introduction

    Calcium carbonate in its pure mineral form is called calcite. In a less pure form it is called limestone. When made from microscopic marine animal shells it is called chalk: Calcium Carbonate = Limestone = Chalk

    Water is acidified in nature by combining with carbon dioxide to form a weak carbonic acid, much weaker than the 5% solution used in this experiment. Hence, natural chemical erosion occurs at a much slower pace.

Answers / Notes

1a.   BEFORE   AFTER

CHALK   WEAK ACID   LID   EROSION PATTERN   RESIDUE

1b. The chalk was chemically dissolved by the weak acid and carried away into solution. The water then evaporated away leaving the eroded residue at the bottom of the lid.

2a. The drop of acid bubbles and fizzes on the surface of the shell.

2b. Shells, like chalk, are made from calcium carbonate. Sea animals take this directly from the ocean water, where it collects from the chemical weathering of limestone.

2c. The shells accumulate as sediment in ocean beds.

3.

SEA ANIMALS   DISSOLVED CALCIUM CARBONATE   CHEMICAL EROSION   LIMESTONE (chalky rock)   CHALK

Materials

☐ The experiment in progress from the previous task card.

☐ A dropper bottle of 5% HCl.

☐ A sea shell. Substitute egg shells if marine shells are not available.

(TO) examine two common cementing agents found in sedimentary rock. To consider the various kinds of rock derived from these cements.

TWO KINDS OF CEMENT   O   Rocks and Minerals (   )

1. Pick out a large transparent quartz crystal from some sand. Fold it inside a small square of paper, then crush it with a pair of pliers.

2. Gather the silica residue in the left corner of a glass slide. Put an equal-sized lump of blackboard chalk in the right corner.

3. Put 1 drop of dilute hydrochloric acid over each mineral. Keep the drops separate.

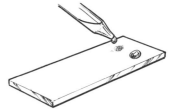

4. Heat to dryness very slowly, moving the slide back and forth over a candle flame. (It splatters if you heat too quickly.) Then let it cool.

5. Draw a picture of the silica residue, labeling the quartz clasts and silica cement. Cite examples of rocks that contain these ingredients.

6. Draw a picture of the chalk residue, labeling the calcite clasts and calcium carbonate cement. Cite examples of rocks that contain these ingredients.

12

Introduction

Clastic sedimentary rock is cemented together by various minerals acting alone or in combination. Your class has already had some experience with calcium carbonate found in shells, blackboard chalk and limestone. Now you need to introduce another common cement called silica (SiO_2).

In crystal form, silica is called quartz. Quartz crystallizes directly from cooling magma to produce many of the crystal flecks you see in granite and other igneous rock. Many gem stones are actually large quartz crystals colored by impurities. Quartz erodes mechanically along with other minerals to form sand. So how do quartz crystals transform into a cementing agent?

Quartz is slightly soluable in water. It dissolves little by little to form both a solution and fine suspension in water. This silica-rich water soaks through loose sediment and gradually cements loose clasts into a solid aggregate. Sand into sandstone is a fine example. Silica also precipitates and hardens in relatively pure form to produce a variety of hard, smooth stones. Agate (found at the beach) or flint (the stuff of arrowheads) are examples of precipitated quartz rock familiar to most students. Petrified wood is another example in which dissolved silica soaks into and replaces wood fiber.

Answers / Notes

5. QUARTZ CLASTS / SILICA CEMENT

Ex: sandstone, agate and flint, petrified wood.

6. CALCITE CLASTS / CALCIUM CARBONATE CEMENT

Ex: chalk, limestone, concrete (in part)

Materials

☐ Coarse brown sand in a pie tin. Several smaller grains of clear quartz can be grouped together in step 1 if larger single grains are unavailable.

☐ A pair of needle nose pliers. The base of the jaws near the fulcrum should have a flat surface area for easy crushing.

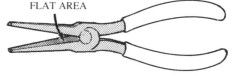

FLAT AREA

☐ A glass slide.

☐ A dropper bottle of 5% HCl.

☐ A candle with drip catcher and matches.

(TO) crack open rocks and save assorted pieces for later identification and display in a rock collection.

CRACKING UP ○ Rocks and Minerals ( )

1. Tape your name on a large tin can. Use it to store the freshly broken rock samples you will collect in step 2.

2. Crack open small rocks between large rocks. CAUTION: Always protect your eyes from sharp flying fragments:
    a. Wear safety goggles,

      (or)

    b. Hold the target rock inside a plastic bag before you strike it.

3. What to keep:
    • Small pieces, about as big as your thumb to the first joint.
    • All kinds of textures and colors.
    • One-of-a-kind pieces. Discard or trade duplicates.

13

Introduction

    Smashing rocks open is wonderful fun, a traditional childhood activity. But flying splinters make this a hazardous venture. Insist that all of your students take measures to protect their eyes at all times. Demonstrate how to contain flying fragments in a plastic bag. (One bag can accommodate perhaps 20 rock-breakings before it becomes too riddled with holes.) These plastic containment bags are not necessary, of course, if you provide protective eye wear.

Answers / Notes

2. This task card might be used on a field trip, for a homework assignment, or in an outdoor class held on a nearby gravel driveway. If you import several large flat "smashing" stones into a corner of your classroom, you can even bring this activity inside for greater supervision. Search out sources of rock in your local environment that offer the greatest possible variety, and can still be safely accessed by students. Then plan accordingly.

Materials

☐ A coffee can or other large, sturdy container.
☐ Masking tape.
☐ Safety goggles or plastic produce bags.
☐ A source of rocks. See note 2 above.

(TO) compare the relative hardness of collected rock samples. To tag them for later identification and analysis.

ROCK HARDNESS Rocks and Minerals (   )

1. You have streak tested minerals. Here's how to streak test a rock.

a. Know the difference between streaking and crumbling…	b. Rub surface to surface. Don't gouge with a point…	c. Test only freshly broken surfaces…

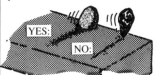

YES: FRESH SURFACE   NO: WEATHERED SURFACE

What kind of rock might crumble against brick yet still scratch?	Make a paper clip (4.0) both streak and scratch a brick (4.5). Explain which is the valid test.	Streaking the outside of a weathered rock might give a false test. Why is this so?

2. Test each rock in your collection can for hardness:
- Tag those that *streak* "st."
- Tag those that *crumble and scratch* "cr."
- Don't tag those that *scratch*.

14

Answers / Notes

1a. The individual clasts or specks of minerals might be hard enough to scratch brick. Yet the cement that binds them together is too weak to hold all the clasts firmly in place. Hence they crumble onto the brick while they scratch. Concrete is a good example of a "rock" that does this.

1b. The paper clip streaks if you draw the *side* of the wire against brick, but scratches if you draw the end of the wire across brick. Since a paper clip (4.0) is actually softer than brick (4.5), streaking with the side of the clip is the valid hardness test.

1c. Minerals on the surface of a rock, exposed to wind and water, first undergo intermediate changes that soften and degrade them before they finally break off the rock or dissolve away. These partially changed minerals may streak across brick even though the unweathered original minerals locked inside still scratch.

Materials

☐ Freshly broken rock samples from each students' collection can.
☐ A brick.

(TO) distinguish between rocks with random and ordered textural patterns. To use these textural clues to identify metamorphic rock.

METAMORPHIC ROCK   O   Rocks and Minerals (   )

1. Intense heat and pressure transforms igneous or sedimentary rock into *metamorphic* (changed) rock. Where on earth might rocks be exposed to such conditions?

2. Sort the rocks in your collection can into these 2 groups:

<table>
<tr><td align="center">FOLIATED
ordered repeating texture</td><td align="center">MASSIVE
random nonrepeating texture</td></tr>
<tr><td align="center"> </td><td align="center"> </td></tr>
</table>

3. a. Look for these metamorphic rocks in your *foliated* group.

> Slate: has hard thin layers; comes from shale.

> Schist: is banded by mica; comes from granite.

b. Look for these metamorphic rocks in your *massive* group.

> Quartzite: has a sugary luster; comes from sandstone.

> Marble: has large white crystals; comes from limestone.

4. Describe in words and pictures each type of rock you are able to classify. (If you find none, describe a friend's example.)

© 1989 by TOPS Learning Systems   15

Introduction

Pass around samples of the 4 rock types listed above for students to observe and handle before you ask them to identify similar rocks in their own collections. Introduce the new vocabulary words "foliated" and "massive" as you talk about each rock.

Answers / Notes

1. Rocks are subjected to increasing heat and pressure the deeper they are buried below the surface of the earth. *(This is known as regional metamorphism.)*

    Rocks exposed to nearby molten magma, in a pluton for example, are changed by intense heat. *(This is known as contact metamorphism.)*

4. *Expect considerable variation among rock samples, especially in color.*

         a. FOLIATED   b. MASSIVE

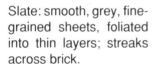   

Slate: smooth, grey, fine-grained sheets, foliated into thin layers; streaks across brick.

Schist: granite-like igneous texture; may have banded streaks of mica; scratches brick.

Quartzite: Massive, tan-colored, translucent, with sugary texture; scratches brick.

Marble: Massive, snow-white, coarse crystallized texture; streaks across brick.

Materials

☐ Samples of slate, schist, quartzite or marble to pass among students as you introduce this activity. (optional)

☐ Freshly broken rocks from each student's collection can.

(TO) begin sorting rocks into 12 basic egg-carton categories on a tentative basis, according to each rock's geological origin.

SORTING ROCKS   O   Rocks and Minerals (   )

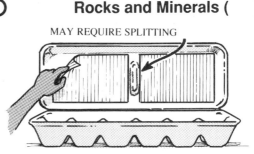

MAY REQUIRE SPLITTING

1. Cut notebook paper to size so it just fits inside an egg-carton lid. Divide it into 12 equal squares.

2. Neatly copy all of this information into the squares, then tape them inside the lid. Number the egg cups in the same order.

1	2	3	4	5	6
CLASTIC SEDIMENTARY coarse grained	CLASTIC SEDIMENTARY fine grained	CLASTIC SEDIMENTARY poorly sorted	CHEMICAL SEDIMENTARY carbonate	CHEMICAL SEDIMENTARY silica	CHEMICAL SEDIMENTARY other
7	**8**	**9**	**10**	**11**	**12**
PLUTONIC IGNEOUS coarse crystals	VOLCANIC IGNEOUS fine crystals	VOLCANIC IGNEOUS pyroclastic	MASSIVE METAMORPHIC quartzite/marble	FOLIATED METAMORPHIC schist/gneiss	METAMORPHIC slate/other

3. Sort rocks from your can into the correct egg cups.
   • If you have no idea how to classify a rock, leave it in your can.
   • If you have some idea where a rock might belong, go ahead and place it where you think it goes. You can always change your mind later.

© 1989 by TOPS Learning Systems   16

Answers / Notes

3. The reason that many rocks were originally collected in a can is so that at least some of them can now be positively identified and classified into this egg-carton collection. It is not necessary to identify each and every rock. Even experts would be hard-pressed to do so.

Remember that this is a rock collection, not a mineral collection. With the exception of certain chemical sedimentary rocks that are made from just a single specific mineral (calcite, silica or coal for example), most minerals do not readily fit into this rock classification scheme.

Materials

☐ Scissors.
☐ An egg carton.
☐ Tape.
☐ Freshly broken rock samples from each student's collection can.

(TO) introduce a six-part write-up procedure for reporting about rocks that students identify over the next four task card activities.

IDENTIFYING ROCKS ◯ Rocks and Minerals (   )

Use the next 4 task cards (in any order) to help identify rocks in your collection. When you are sure about a rock, fill out an ID card like this one, using quarter pieces of notebook paper...

3. clastic sedimentary, poorly sorted:
CONCRETE

gravel clasts
sand clasts
carbonate cement matrix

◀ 1. egg cup category
◀ 2. name
◀ 3. labeled drawing
◀ 4. tests
◀ 5. description
◀ 6. formation

TESTS: Fizzes with dilute acid; crumbles across brick.

DESCRIPTION: A conglomerate of rounded sand and gravel clasts cemented in a carbonate matrix.

FORMATION: A human-made rock. Probably once part of a sidewalk or building foundation.

...Continue in this manner until you classify from 1 to 3 rocks per egg cup, at least 24 rocks in all. This will require several days of thoughtful work. During this time keep looking for rocks to fill empty positions, or trade with friends.

17

Introduction

The 4 task cards that follow this one provide enough general information to enable your students to correctly identify many rocks. Depending on where you live, certain categories in 1-12 may have an overabundance of rocks while other categories have none at all. This is a good time for students to trade for rocks they still need, and search for new ones that fit particular sets of criteria.

Once a rock is identified with certainty, students should complete its ID card. The first 5 steps are descriptive, and should be completed without difficulty. Step 6, explaining the geological formation of each rock, requires more sophisticated hypothesizing. To succeed here, your class needs to thoroughly understand the teaching notes found *below* each card. This information is best covered in a lecture/discussion format. Briefly explain how each rock forms while students take appropriate notes. Tell students that you expect to get this information back again, as they apply similar ideas to specific rocks in their collections.

Allow students perhaps 4 class periods to classify the required minimum of 24 rocks. Rock hounds will want to work beyond this minimum because of their natural interest in the subject. Others may be motivated to search for more rocks and write additional cards if you offer extra credit. Once students understand the required format, these rock collections offer a great opportunity for independent study. Over the next 3 weeks or so before this module ends, your students will have ample time, both at school and at home, to assemble a grand egg-carton collection of rocks, each with its own ID card.

Check out all available rock and mineral field guides from your school library and keep them in class for student reference. These guides will carry students beyond the bare essentials presented here.

Materials

☐ An egg carton collection of freshly broken rocks, plus extras in a can.
☐ Quarter pieces of notebook paper.
☐ Scissors (optional.)

(TO) identify and describe clastic sedimentary rocks that belong to egg-cup categories 1-3.

CLASTIC SEDIMENTARY (1-3) ○   Rocks and Minerals ( )

1	2	3
coarse grained	**fine grained**	**poorly sorted**

1
coarse grained

SANDSTONE
coarse   medium

always:
• clasts have uniform size (well sorted)
• clasts are big enough to see individually
usually:
• rough to touch (unless worn smooth)
• cemented by silica
sometimes:
• crumbles across brick
• scratches brick
occasionally:
• contains fossils, water ripples
• has bedded layers

2
fine grained

CLAY OR SILTSTONE

SHALE

always:
• smooth to touch
• clasts too small to see individually
usually:
• cemented by clay
• streaks across brick
• breaks along flat horizontal surfaces
sometimes:
• has bedded layers
occasionally:
• contains fossils, water ripples

3
poorly sorted

CONGLOMERATE

ORE BEARING

always:
• poorly sorted
usually:
• large and small clasts (poorly sorted sizes)
• color variations (poorly sorted minerals)
• crumbles or scratches across brick
sometimes:
• cemented by carbonates, silica, or clay; if rust colored, probably cemented by iron
• bubbles with dilute acid

18

Introduction

1
MEDIUM SANDSTONE
The clasts in this rock have been transported over long distances by wind and water because they are eroded to small, rounded pieces that are well sorted to a uniform size. Quartz is the predominant grain. Mica and feldspar, which are less stable minerals, have long since eroded away. (This is often the case with beach sand.) The rock scratches brick without crumbling, indicating it is strongly cemented by dissolved silica.

COARSE SANDSTONE
The clasts in this rock have consolidated much more rapidly. They are larger, more angular, and not as well sorted. Mica and feldspar are still well represented among the quartz grains. This sediment was evidently transported over just a short distance before being deposited in a lake or river bed.

2
SILT STONE
The flat bedding plane indicates deposition in a marine environment. Ripple patterns on the surface suggest gentle shaping by slow-moving water.

SHALE
Very fine clay particles must have settled out over a long period of time to create the flat, smooth, uniform, grey texture of this rock. Conditions in deep ocean water are calm enough for this to occur. Seasonal variations in the suspended clay particles might account for different shades of grey among the layers that form the bedded plane.

3
CONGLOMERATE
Consolidation of random gravel sizes indicates a turbulent, chaotic sedimentation process, possibly associated with glacial activity. A positive acid test reveals some cementation by calcium carbonate. The cementation is weak enough to allow crumbling as the rock is scratched across brick.

ORE BEARING
Streaks of iridescent black, red and yellow indicate iron compounds; streaks of light yellow indicate aluminum compounds. The texture is earthy and crusty, suggesting a gradual decomposition and oxidation of metals as they leach by chemical erosion from the stone.

(TO) identify and describe chemical sedimentary rocks that belong to egg-cup categories 4-6.

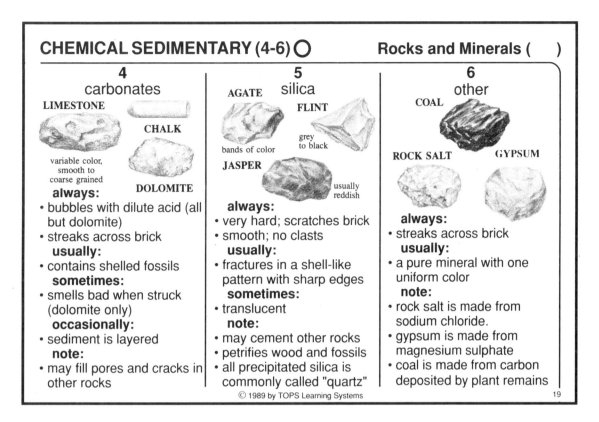

CHEMICAL SEDIMENTARY (4-6) ○   Rocks and Minerals (   )

4
carbonates
LIMESTONE

CHALK

variable color,
smooth to
coarse grained

DOLOMITE

always:
• bubbles with dilute acid (all but dolomite)
• streaks across brick
usually:
• contains shelled fossils
sometimes:
• smells bad when struck (dolomite only)
occasionally:
• sediment is layered
note:
• may fill pores and cracks in other rocks

5
silica
AGATE

FLINT

grey to black

bands of color

JASPER

usually reddish

always:
• very hard; scratches brick
• smooth; no clasts
usually:
• fractures in a shell-like pattern with sharp edges
sometimes:
• translucent
note:
• may cement other rocks
• petrifies wood and fossils
• all precipitated silica is commonly called "quartz"

6
other
COAL

ROCK SALT

GYPSUM

always:
• streaks across brick
usually:
• a pure mineral with one uniform color
note:
• rock salt is made from sodium chloride.
• gypsum is made from magnesium sulphate
• coal is made from carbon deposited by plant remains

© 1989 by TOPS Learning Systems   19

Introduction

4

LIMESTONE
Fizzes with dilute acid, indicating the presence of calcium carbonate. Marine animals extract this from sea water to make their shells. These shells accumulate on the seabed and compact through deep burial. Shell fossils are common throughout if they haven't been crushed beyond recognition.

CHALK
Fizzes with dilute acid. This calcium carbonate consolidates from the accumulation of microscopic shelled marine animals in the manner of limestone.

DOLOMITE
Forms originally like limestone. Then the calcium is gradually replaced by the action of sea water to form a carbonate of magnesium. Because it contains less calcium, dolomite does not readily fizz with dilute acid.

5

AGATE
(Also called chalcedony.) Chemical weathering of quartz crystals creates a fine suspension of silica in sea water that gradually settles out and hardens. As chemical conditions change, the color of the suspension and its rate of deposition change as well, forming banded variations.

FLINT
(Also called chert.) Diatoms and other microscopic marine animals remove silica from sea water to form their skeletal structures. These collect on the seabed after the animals die, then compress and harden through deep burial. (Compare to limestone formation.)

JASPER
(Also called chert.) Silica precipitates from sea water at great depths together with other materials, usually red iron oxide.

6

COAL
The uniform black color indicates that coal is made from pure carbon, the organic remains of plants. Its banded appearance suggests that plant material was deposited layer upon layer in a swampy environment that prevented normal biological decay.

ROCK SALT
The uniform white to colorless crystals with perfect cubic cleavage are characteristic of table salt. Large crystals suggest slow growth. They likely formed in the bed of an evaporating desert lake or inland sea.

GYPSUM
Crystalline flakes of calcium sulphate may be white, grey, red, green or brown due to various impurities. Gypsum precipitates, like rock salt, as desert lakes and inland seas evaporate. Used to make Plaster of Paris.

notes 19

(TO) identify and describe igneous rocks that belong to egg-cup categories 7-9.

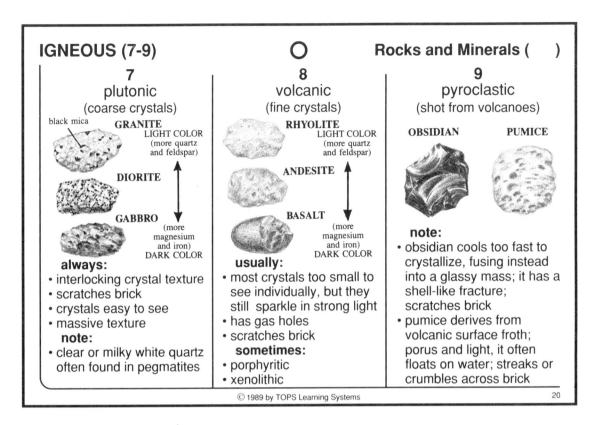

IGNEOUS (7-9)　　　　　○　　　Rocks and Minerals (　)

7
plutonic
(coarse crystals)

GRANITE
black mica
LIGHT COLOR
(more quartz
and feldspar)

DIORITE

GABBRO
(more
magnesium
and iron)
DARK COLOR

always:
• interlocking crystal texture
• scratches brick
• crystals easy to see
• massive texture
　note:
• clear or milky white quartz
  often found in pegmatites

8
volcanic
(fine crystals)

RHYOLITE
LIGHT COLOR
(more quartz
and feldspar)

ANDESITE

BASALT
(more
magnesium
and iron)
DARK COLOR

usually:
• most crystals too small to
  see individually, but they
  still sparkle in strong light
• has gas holes
• scratches brick
　sometimes:
• porphyritic
• xenolithic

9
pyroclastic
(shot from volcanoes)

OBSIDIAN　　**PUMICE**

note:
• obsidian cools too fast to
  crystallize, fusing instead
  into a glassy mass; it has a
  shell-like fracture;
  scratches brick
• pumice derives from
  volcanic surface froth;
  porus and light, it often
  floats on water; streaks or
  crumbles across brick

© 1989 by TOPS Learning Systems　　　　20

Introduction

7

Vocabulary Note: Igneous rocks with very large crystals (over 1 cm) are called "pegmatites." Granite pegmatites are prized for their large, beautiful crystals.

GRANITE
Because its crystals are large enough to see individually, this rock must have solidified gradually from slowly cooling magma in an underground pluton. Its light color indicates the rock is rich in quartz and feldspar crystals.

DIORITE
This coarse-grained rock has a geological history similar to granite. Its overall gray color indicates a higher concentration of dark colored minerals with less quartz and feldspar.

GABBRO
This greenish-gray rock has large crystals rich in magnesium and iron. These minerals solidify early from magma, at higher temperatures than lighter colored quartz and feldspar. Hence, gabbro forms deep in the earth, far below granite.

8

Vocabulary Note: Igneous rocks with larger crystals scattered throughout a fine-grained matrix are called "porphyritic." Igneous rocks with unmelted rock fragments caught in the magma are called "xenolithic."

RHYOLITE
A light-colored rock of granitic composition. Its porphyritic texture suggest the magma cooled slowly underground then erupted to finish cooling rapidly at the surface.

ANDESITE
This rock is found in regions of great tectonic activity. It is a grand mix of colliding plates that partially remelt at great depths and then erupt to the surface. As a result it is sometimes both porphyritic and xenolithic.

BASALT
The dark color indicates high magnesium and iron content. As these minerals oxidize, basalt changes to brown or red. It erupts from deep in the earth to form vast surface lava flows. Expanding bubbles of gas mark basalt with numerous pores.

9

OBSIDIAN
The rock is fused into a hard and smooth glassy mass because its solidifying magma cooled too rapidly to form crystals. This implies that it was air-cooled right at the surface of a volcano, or that viscous fragments of lava were tossed high into the air causing immediate solidification.

PUMICE
This rock was probably expelled violently from a volcano in the initial gas-rich explosive phase. It resembles a kind of frozen volcanic froth.

(TO) identify and describe chemical metamorphic rocks that belong to egg-cup categories 10-12.

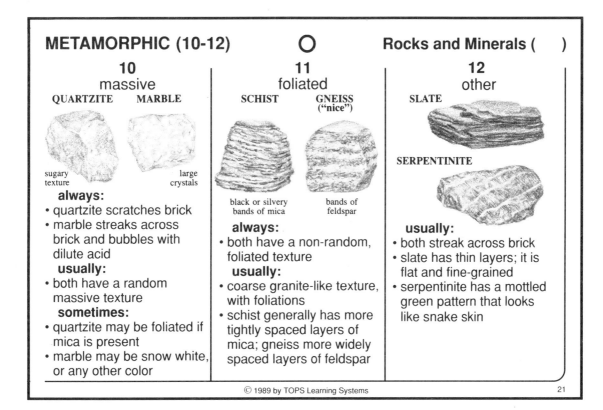

METAMORPHIC (10-12)   O   Rocks and Minerals (   )

10
massive
QUARTZITE   MARBLE

sugary texture   large crystals

always:
• quartzite scratches brick
• marble streaks across brick and bubbles with dilute acid
usually:
• both have a random massive texture
sometimes:
• quartzite may be foliated if mica is present
• marble may be snow white, or any other color

11
foliated
SCHIST   GNEISS ("nice")

black or silvery bands of mica   bands of feldspar

always:
• both have a non-random, foliated texture
usually:
• coarse granite-like texture, with foliations
• schist generally has more tightly spaced layers of mica; gneiss more widely spaced layers of feldspar

12
other
SLATE

SERPENTINITE

usually:
• both streak across brick
• slate has thin layers; it is flat and fine-grained
• serpentinite has a mottled green pattern that looks like snake skin

21

Introduction

10

QUARTZITE
Quartz sandstone was buried at great depth where it was subjected to high temperature and pressure. This fused and melted the individual sand grains. In their place new quartz crystals intergrew and enlarged, forming a hard crystal matrix that looks somewhat like sugar.

MARBLE
A soft streak across brick, as well as a positive acid test, reveals that marble is made from calcite (calcium carbonate). It formed from limestone that was buried at great depth until heat and pressure recrystallized the calcite into coarse crystals.

11

SCHIST
This rock metamorphosed from mica-rich clays, sand or granite. Mica is a flat and platy mineral that crystallizes perpendicular to the direction of greatest force. When buried deep in the earth, it forms silvery foliations of white mica or dark foliations of black mica that run horizontal to the vertical compression.

GNEISS
As schist is subjected to even greater temperatures and pressures, its narrow bands of mica metamorphose into broader bands of feldspar. Gneiss (pronounced "nice") not only contains more feldspar than schist, it also breaks with difficulty into coarser pieces, while mica schist fractures easily into thin units.

12

SLATE
This rock is the first to metamorphose from sedimentary shale. As such, it is less fragile, but still soft enough to streak across brick. With increasing heat and pressure, mica begins to form, foliating the slate to schist. Further metamorphosis changes schist to gneiss.

SERPENTINITE
Magnesium and iron-rich basalt degrade at temperatures lower than those forming the original igneous rock to produce a green fibrous mineral called chlorite. Chlorite (also known as green mica) gives serpentinite its characteristic snake-skin appearance. Related forms are greenschist and amphibole, also both green.

(TO) understand how igneous, sedimentary and metamorphic rocks continuously change form, creating one grand cycle of change. To see the bigger picture.

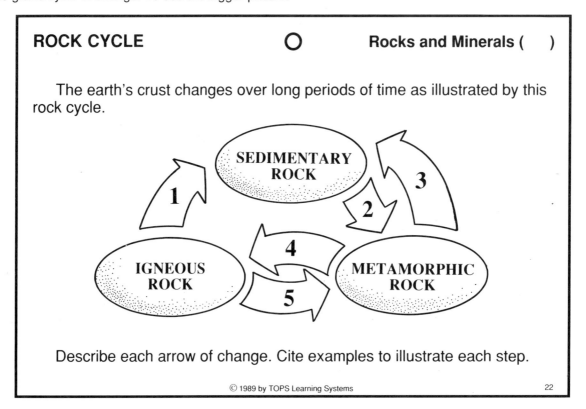

ROCK CYCLE   O   Rocks and Minerals (   )

The earth's crust changes over long periods of time as illustrated by this rock cycle.

Describe each arrow of change. Cite examples to illustrate each step.

22

Answers / Notes

1. Igneous rock (like granite) both weathers away mechanically and dissolves away chemically. Its rock clasts are sorted and transported by wind and water, deposited layer upon layer, and gradually cemented together by dissolved minerals (like silica) to form sedimentary rock (like sandstone).

2. Sedimentary rock (like shale) gets buried ever more deeply as new sediment piles up on top of old. Pressure and heat gradually build, transforming it into new rock forms (first slate, then schist, then gneiss).

3. Metamorphic rock (like marble) is uncovered by surface erosion. A river, for example, cuts out a deep gorge. Exposed once again to surface weathering, the rock erodes away chemically (as dissolved calcium carbonate), is carried off to the sea, assimilated by marine animals, and redeposited as shells to eventually form new sedimentary rock (limestone).

4. Metamorphic rock (like quartzite) that is buried ever more deeply will sooner or later melt along with its surrounding rock into molten magma. This magma mixes together, then slowly cools in underground plutons (forming igneous granite rock) or erupts in surface lava flows (forming igneous basalt rock).

5. Igneous rock (like granite) that solidifies in underground plutons may never reach the surface. Instead it may be pushed lower, subjected to increasing heat and pressure which changes it into new forms (first schist, then gneiss).

Materials

None.

(TO) represent the relative abundance of common rocks as a bar graph. To develop an overview.

ORDINARY ROCK ⭕ Rocks and Minerals (   )

Over 99% of the earth's crust is made from the following rocks:

dark igneous (mostly basalt) = 42.7%	marble = 0.9%
gneiss = 21.4%	sandstone = 1.7%
light igneous (mostly granite) = 22.0%	schist = 5.1%
limestone = 2.0%	shale = 4.2%

1. What percent of the earth's crust is made from…

    a. sedimentary rock?
    b. metamorphic rock?
    c. igneous rock?

2. Represent all 8 ordinary rocks on a bar graph. Arrange them in this order: sedimentary – metamorphic – igneous.

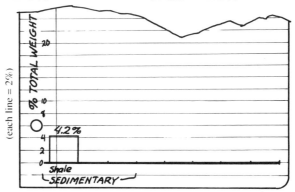

3. Some of these rocks don't seem to be all that common; at least they were hard to find for your rock collection. Explain this discrepancy.

© 1989 by TOPS Learning Systems   23

Answers / Notes

1.   a. SEDIMENTARY ROCK

    limestone = 2.0%
    sandstone = 1.7%
    shale = <u>4.2%</u>
         7.9%

   b. METAMORPHIC ROCK

    gneiss = 21.4%
    marble = 00.9%
    schist = <u>05.1%</u>
       27.4%

   c. IGNEOUS ROCK

    basalt plus = 42.7%
    granite plus = <u>22.0%</u>
         64.7%

2.

3. This graph represents all material in the earth's crust. But rock collectors can only access what is available at the earth's surface, within the locality that they are searching. *(Gneiss, for example, comprises about a fifth of the earth's crust. But only a tiny fraction of that amount is found at the earth's surface, and even that is not well distributed or easily accessible.)*

Materials

☐ Lined notebook paper.
☐ A ruler or straight edge.

(TO) compare the permeability of chalk and clay. To evaluate which types of collected rocks are the most water permeable.

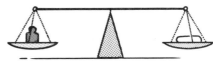

PERMEABILITY

1. Get a piece of chalk and some oil-based clay about as big as your thumb. Find the mass of each to the nearest tenth of a gram.

2. Touch a single drop of water to each substance.
   a. Which material has the greatest permeability (allows water to soak through most easily)?
   b. Predict which material will gain the most mass if you soak both in water.

   CLAY   CHALK

3. Soak the chalk and clay in a jar of water for about 1 minute.
   a. Write your observations.
   b. Dry each substance with a paper towel, then reweigh to the nearest tenth of a gram. Evaluate your prediction.

4. Find the most permeable rock in your collection. What kind of rock is it? Does your answer make sense? Explain.

© 1989 by TOPS Learning Systems   24

Answers / Notes

1. Answers will vary by sample size.

2a. The water drop beaded on the clay but instantly soaked into the chalk. Chalk, therefore, has much greater permeability to water.

2b. The chalk should gain considerable mass as water soaks into it. The clay should gain little or no mass because it is impermeable to water.

3a. Chalk bubbles vigorously as water soaks in and displaces the air inside it, but clay doesn't bubble at all.

3b. Answers will vary by sample size. Here is one result:

	CHALK	CLAY
after soaking	7.9 g	32.4 g
– before soaking	5.4 g	32.4 g
water absorbed	2.5 g	0.0 g

As predicted, the chalk gained considerable mass, while the clay gained none at all.

4. *(Students should add a drop of water to each rock in their collection and observe how fast it soaks in).* Any soft sedimentary rock (including chalk) should absorb water most readily, as well as pyroclastic pumice. These rocks are soft and porous, presenting no barrier to the passage of water.

Materials
☐ A short piece of blackboard chalk.
☐ A thumb-sized piece of oil-based clay.
☐ A gram balance.
☐ A jar or beaker of water.
☐ An eyedropper.
☐ A paper towel.

(TO) determine the density of clay. To discover that each substance has a unique density, regardless of the volume measured.

DENSITY   O   **Rocks and Minerals (   )**

1. Fill a graduated cylinder with exactly 50.0 ml of water. Add clay pebbles until the water level raises to 60.0 ml. If you overshoot, try again.
   a. What volume of clay is in the cylinder?
   b. Remove the clay from the water, dry the pebbles and press them together. Find the mass of this lump to the nearest tenth gram.

2. Density is defined as the mass of any substance divided by its volume.

$$D = \frac{mass}{volume} = \frac{g}{ml}$$

   a. Find the density of your clay ball.
   b. In a similar manner, make a 20 ml ball of clay and find its density.
   c. Make a 5 ml ball of clay and find its density.

3. How do the densities of each clay ball compare? Make a generalization about density.

25

Answers / Notes

1a. volume of clay = 60.0 ml - 50.0 ml = 10.0 ml
1b. mass of clay   = 16.7 g

2a. density of 10 ml clay = 16.7 g / 10 ml = 1.67 g/ml
2b. density of 20 ml clay = 33.6 g / 20 ml = 1.68 g/ml
2c. density of 5 ml clay = 8.2 g / 5 ml = 1.64 g/ml

3. Within the limits of measuring error, each clay ball has the same density. This implies that density is constant, regardless of how much matter you choose to measure. *(Another way to understand this is to consider a property of fractions: a small mass divided by a small volume can equal the same quotient as a large mass divided by a large volume.)*

Materials

☐ A 100 ml graduated cylinder.
☐ A thumb-sized piece of oil-based clay.
☐ A gram balance.
☐ A paper towel.

(TO) compare the densities of granite and basalt. To relate density to the position of rocks within the earth's crust.

DENSITY OF IGNEOUS ROCK ⃝   Rocks and Minerals ( )

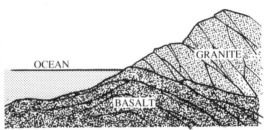

1. Cut a circle of paper that *just* fits inside your graduated cylinder.

2. Use this circle as a template to find 4 pieces of granite that are reasonably large, but still fit inside your cylinder.
   a. Weigh these 4 pieces of granite together, to the nearest tenth gram.
   b. Find the volume of all 4 pieces in your graduated cylinder.
   c. Divide mass by volume to find the density of granite.

3. In a similar manner, find the density of four pieces of basalt.

4. Ocean basins are composed mainly of basalt; the continents of granite and related rocks. Use your density results to propose a theory that might explain why this is so.

© 1989 by TOPS Learning Systems   26

Introduction

Pressures deep inside the earth's crust are high enough to squeeze solid rock so it acts like a slow-moving liquid. This allows rocks in the earth's crust to move up or down. Ilustrate this by pouring a few tablespoons of salt and pepper into a small jar or beaker. Gently shake the beaker, showing your class how "pepper rocks" migrate to the top and "salt rocks" sink to the bottom. Discuss how the density of each kind of "rock" determines its overall movement.

Answers / Notes

Granite has a slightly lower density than basalt. Masses and volumes will vary widely, but the calculated densities should closely match these values.

2. GRANITE
a. mass = 24.5 g
b. volume = 9.2 ml
c. density = 24.5 g / 9.2 ml = 2.7 g/ml

3. BASALT
a. mass = 15.1 g
b. volume = 5.2 ml
c. density = 15.1 g / 5.2 ml = 2.9 g/ml

4. Granite, being a less dense rock, works its way to the top of the earth's crust. Basalt, being more dense, settles in at greater depth. *(This theory that the earth's crust is somewhat plastic and that rocks float in equilibrium is called isostasy.)*

Materials

☐ Table salt and pepper.
☐ A small beaker or jar.
☐ A pair of scissors.
☐ A 100 ml graduated cylinder.
☐ Small pieces of granite and basalt.
☐ A gram balance.
☐ A calculator (optional).

(TO) account for differences in density between crystallized silica and fused silica.

DENSITY OF SILICA   O   Rocks and Minerals (   )

1. Use a penny to draw circles
on paper . . .

   a. Arrange 15 circles into a tight,   b. Arrange another 15 circles into a
organized pack.   loose, disorganized pack.

   MINIMUM
SPACE
      RANDOM
SPACE

2. The silica in agates crystallizes in a slow, organized way. The silica in glass
marbles fuses quickly into a disorganized mass.

   a. How would you expect the density of glass marbles to compare with the
density of agates? Use your drawings in step 1 to support your hypothesis.
   b. Test your hypothesis. Take measurements and calculate densities.

© 1989 by TOPS Learning Systems   27

Answers / Notes

1a.
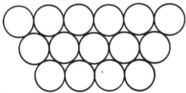
      1b.
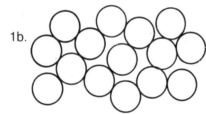

2a. An ordered arrangement of precipitated and crystallized silica represented by drawing (a) above should occupy less volume than the more disordered arrangement of fused silica represented by drawing (b). Agates, therefore, should have a higher density than glass marbles. *(This explanation is somewhat oversimplified. The silica in glass marbles is mixed with soda and lime to lower its melting point and make the glass easier to work. These relatively lighter compounds also serve to lower the density of glass marbles.)*

2b. *Students should compare the densities of agates with glass marbles in a manner similar to the previous task card. Igneous quartz can be logically substituted for agate. It, too, crystallizes slowly, not from precipitated silica, but from cooling magma. Glass marbles should have a density that is slightly less than either form of quartz. Here is a sample calculation.*

AGATES	GLASS MARBLES
mass = 26.7 g	mass = 30.1 g
volume = 10.2 ml	volume = 11.9 ml
density = 2.6 g/ml	density = 2.5 g/ml

As predicted, the more highly organized agates had a slightly higher density than the less organized fused glass.

Materials

☐ A penny.
☐ Silica rocks (agate or quartz), small enough to fit inside a 100 ml graduated cylinder.
☐ Glass marbles.
☐ A gram balance.
☐ A calculator (optional).

(TO) compute the density of water. To use this special result to develop the notion of specific gravity.

SPECIFIC GRAVITY ○ Rocks and Minerals ( )

1. Find the density of 100 ml of water. Show your work.
   a. Check your answer by finding the density of 50 ml of water.
   b. The density of water is always very close to what whole number?

2. *Specific gravity* is the density of any substance divided by the density of water.

$$S.G. = \frac{\text{density of substance}}{\text{density of water}}$$

   a. What is the specific gravity of granite? Of basalt?
   b. How does the specific gravity of a substance compare with its density?

3. Pumice often floats on water. What can you conclude about the specific gravity of this rock?

4. The whole earth, on average, is 5.5 times denser than water. (It has a specific gravity of 5.5.)
   a. Compare this with the specific gravity of rocks in the earth's crust.
   b. How is matter near the center of the earth different from matter near its surface?

© 1989 by TOPS Learning Systems

28

Answers / Notes

1. The density of water is always very close to 1 g/ml. *(Our metric system was defined that way!)*

100 ml water:	50 ml water:
cylinder and water =167.6 g	cylinder and water =117.6 g
(minus) cylinder = 67.6 g	(minus) cylinder = 67.6 g
water =100.0 g	water = 50.0 g

$$\text{density} = \frac{100.0 \text{ g}}{100 \text{ ml}} = 1 \text{ g/ml} \qquad \text{density} = \frac{50.0 \text{ g}}{50 \text{ ml}} = 1 \text{ g/ml}$$

2a. specific gravity of granite = 2.7 g/ml ÷ 1.0 g/ml = 2.7
    specific gravity of basalt = 2.9 g/ml ÷ 1.0 g/ml = 2.9

2b. Specific gravity has the same numerical value as density. The only difference involves units. Density has units of g/ml. Specific gravity, being the ratio of 2 numbers, has no units at all.

3. If pumice floats on water, it must be less dense than water. Thus, its specific gravity must be less than 1.

4a. The specific gravity of rocks measured thus far ranges between 2.6 and 2.9. On average, this is only about half the specific gravity of the earth taken as a whole.

4b. Matter near the center of the earth is about twice as dense as surface rock. (The earth's crust is extremely thin relative to its vast interior. Hence, the density of rock has an insignificant effect on the earth's grand average — much like a fly resting on a brick.

Materials

☐ A 100 ml graduated cylinder.
☐ A gram balance.

enrichment

(TO) understand how freezing water contributes to the mechanical erosion of rock.

ICE AND WATER Rocks and Minerals (   )

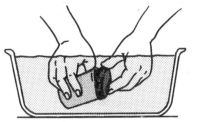

1. Gently dunk a film canister in a tub of standing water without stirring it up. Snap on the cap while still under water so no air pockets remain inside.
2. Freeze it overnight.

   a. As ice freezes, does its volume increase or decrease? How do you know?
   b. Predict what happens to the film canister after the water freezes.
   c. Wait 24 hours to complete the rest of this task card.
3. Measure the ice and canister with a mm ruler. Calculate the % increase in the length of the ice.

$$\% \text{ increase} = \frac{\text{extra length}}{\text{original length}} \times 100$$

4. Slide the ice out of its canister. Draw a diagram to show where dissolved air and ice cracks form in the ice.
5. Freezing and melting water is the main cause of mechanical rock erosion. Explain how this works.

29

Answers / Notes

2a. Ice expands into a greater volume as it freezes . This lowers its density, causing ice to float on water.
2b. The ice inside the fixed volume of the canister will expand, popping off the snap-on cap.

3.   length of ice sticking out = 4.3 mm
          length of film canister = 48.5 mm
                    % increase = 4.3 mm / 48.5 mm x 100 = 9%

4. *Insist on an accurate drawing. It provides the starting point for activity 31.*

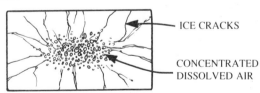

ICE CRACKS

CONCENTRATED DISSOLVED AIR

5. Water seeps into fractures in the surface of a rock and then it freezes. The expanding ice mechanically widens some cracks and possibly splits off fragments. This process repeats through each new thaw and freeze cycle until, after many years, the rock mechanically erodes away to rubble.

Materials

☐ An empty film canister with snap on lid.
☐ A tub of standing water. Better contrast is obtained if concentrations of dissolved air are reduced by allowing the water to stand overnight.
☐ A freezer. Water samples might be tagged with student names, then collected at the end of the period to be transported in one lot to the school cafeteria for freezing.
☐ A calculator (optional).

(TO) study how the concentration of salt affects the size of crystals that precipitate from solution.

CONCENTRATION AND CRYSTAL SIZE

O   **Rocks and Minerals ( )**

1. Divide a lid or glass pane into 4 parts. Label them with these letters:

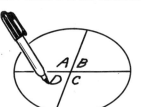

2. Add the given number of saturated *salt water* drops to each section with an eyedropper.

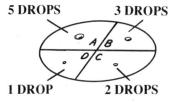

5 DROPS   3 DROPS

1 DROP   2 DROPS

3. Now dilute these salty drops by adding the given number of *fresh water* drops to each section.

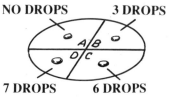

NO DROPS   3 DROPS

7 DROPS   6 DROPS

4. Find the % of salt water saturation in each puddle:

$$\% = \frac{\text{salt water drops}}{\text{total drops}} \times 100$$

5. Predict the relative sizes of salt crystals that remain after the water evaporates from each puddle. Explain your reasoning.

6. Evaluate your prediction after 24 hours. Report, in words and pictures, how mineral concentration relates to crystal size.

© 1989 by TOPS Learning Systems   30

Answers / Notes

4.   A. 5 drops / 5 drops x 100 = 100 %
      B. 3 drops / 6 drops x 100 = 50%
      C. 2 drops / 8 drops x 100 = 25%
      D. 1 drop   / 8 drops x 100 = 12.5%

5. Crystal size should increase in direct proportion to salt concentration. Thus puddle (A) with a 100 % concentration should have the largest crystals; puddle D with 12.5% concentration should have the smallest crystals.

6.

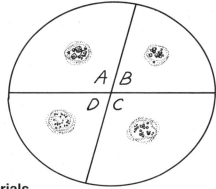

Higher salt concentrations did, in fact, produce larger crystals; lower salt concentrations produced smaller crystals. *(These results are helpful in understanding the formation of pegmatites in activity 31.)*

Materials

☐ A flat surface made from painted metal, plastic or glass. Coated canning jar lids, margarine lids, or square panes of glass work well.
☐ A saturated solution of salt water.
☐ An eyedropper.
☐ A jar of fresh water.

(TO) explain how veins of pegmatite form in plutons of granite.

PEGMATITES ○ Rocks and Minerals (   )

A freezing film canister of water closely models a cooling pluton of granite. Study these similarities, then answer the 3 questions below.

FREEZING WATER CANISTER   **COOLING GRANITE PLUTON**

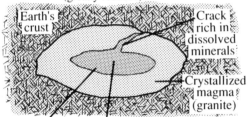

1. Dissolved air doesn't easily freeze in water. How does this fact explain why air concentrates in the center of the ice, as it did in activity 29?
2. Water vapor, quartz, mica, feldspar, tourmaline, beryl and topaz all tend to concentrate in the center of cooling granite plutons. What do these minerals share in common with dissolved air?
3. Granite cracks as it solidifies and cools. Use results from activity 30 to explain why these cracks fill with extra large crystals, called pegmatites.

31

Answers / Notes

1. The film canister tends to cool fastest on the outside, retaining heat longer near the center. Thus ice freezes from outside in, driving higher and higher concentrations of air (which doesn't easily freeze in ice) towards the center.

2. Like air, these minerals are the most volatile — the last to crystallize from liquid magma. As granite plutons solidify from the outside in, these volatiles are driven toward the liquid center in higher and higher concentrations, just like air in solidifying water.

3. Mineral-rich molten magma trapped in the center of the pluton flows into cracks that open into the surrounding granite that is cooling and contracting. The previous task card demonstrated the link between mineral concentration and crystal size. Because this magma is mineral rich, crystals of exceptional size and beauty solidify in the cracks, forming veins of pegmatite.

Materials

☐ Nothing is required other than the results from task cards 29 and 30.

(TO) soak a paper towel in Epsom salts and speculate how this might "petrify" the paper.

PETRIFIED PAPER   ○   **Rocks and Minerals (   )**

1. Fill a small jar half full with Epsom salts (magnesium sulfate crystals). Fill with water and stir well.

2. Bend a paper clip and tape it to the jar like this. Set the jar on a wide plastic lid.

3. Cut a paper towel in 2 halves. Crumple one into a ball; fold the other into a narrow strip. Soak both in the Epsom salts.

SOAKED TOWELS

4. Put the strip in the jar with one end resting on the paper clip. Rest the crumpled paper on the jar's rim. Set this experiment aside to use in the next activity.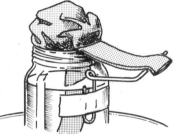

5. Wood petrifies (turns to stone) as dissolved silica replaces wood fiber. Predict how magnesium sulfate might "petrify" this paper towel in a similar manner.

32

Answers / Notes

4. *Students may proceed directly into the next activity, after they complete step 5. Or they may continue on another day.*

5. The magnesium sulfate may not dissolve away or replace the paper fiber, but it will probably soak into air spaces between the fibers and then crystallize, giving the soft paper towel a hard, stone-like quality.

Materials
- ☐ Epsom salts (magnesium sulfate).
- ☐ A small jar or beaker.
- ☐ Water.
- ☐ A stirring rod (pencil).
- ☐ A paper clip.
- ☐ Masking tape.
- ☐ A wide plastic margarine lid or equivalent.
- ☐ A paper towel.

(TO) model how silicates petrify wood and how carbonates form stalactites and stalagmites in caves.

CHEMICAL ICICLES   ○   Rocks and Minerals (   )

1. Set your "petrifying" paper in a warm dry place. *After* it is totally dry, answer these questions.

a. How did the paper change?

b. Will it burn in a candle flame?

c. How is this paper similar to petrified wood? Evaluate your prediction.

2. Adjust the paper strip so the end hangs over the paper clip, slightly below the level of solution in the jar. It should drip *very* slowly.

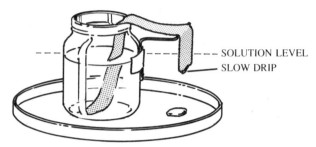

— SOLUTION LEVEL
— SLOW DRIP

a. How do stalactites and stalagmites form in caves? What are they made from? (Use a dictionary if you need one.)

b. See if you can grow these formations from magnesium sulfate over the next few days. Report your results in words and pictures.

33

Answers / Notes

1a. The paper towel has become hard and inflexible, retaining its original crumpled shape.

1b. The magnesium sulfate melts and bubbles, turns brittle and starts to crumble, but the paper will not burn. It only turns slightly black around the edges.

1c. As predicted, magnesium sulfate soaked into the paper and hardened. In a similar manner (though not identical), silica replaces wood and turns to stone. Neither paper nor wood burn once this happens.

2a. Stalactites are deposits of calcium carbonate that form like icicles as water drips from ceilings of limestone caves. Stalagmites form in a similar manner, only they grow upward from the floor as they are coated with mineral rich drops that drip from the stalactites above.

2b.

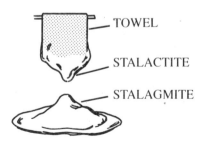

— TOWEL
— STALACTITE
— STALAGMITE

A small stalactite first begins to form on the end of the towel followed by a stalagmite lump that begins to grow up from the lid. *(Humidity needs to be somewhat low for this to occur. At first, solution will flood and crystallize over the lid. But as the drip slows down, first the stalactite and then the stalagmite begin to grow. If conditions are real dry, the two formations may even join. The crystals are very fragile, however, because they are initially hollow. They may not span more than several centimeters or so before crumbling under their own weight. It may be necessary to adjust the end of the towel downward from time to time to renew the drips.)*

Materials

☐ The Epsom salts experiment in progress from the previous activity.
☐ A candle with drip catcher and matches.
☐ A dictionary.

(TO) model a silica tetrahedron. To demonstrate how this negative unit combines with positive ions to form neutral minerals.

THE SILICA TETRAHEDRON  Rocks and Minerals (   )

1. If oxygen is drawn as large as a penny, here's how the 8 basic elements compare as ions:

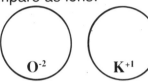  

O^{-2} K^{+1} Ca^{+2} Na^{+1} Mg^{+2} Fe^{+2} Al^{+3} Si^{+4}

2. Roll 4 clay "oxygen" balls to the diameter of a penny. Fit these around 1 tiny "silicon" ball to model a *silica tetrahedron.*

a. Why do chemists call these atoms "SiO_4"? Add up their net charge.

b. Minerals that contain these tetrahedra are called "silicates". They are always neutral. Show that $Mg_3Al_2(SiO_4)_3$ is a neutral silicate.

3. Silicates comprise 92% of all rock-forming minerals!

a. Complete this table.
b. Show that each silicate is neutral.

(Save your model.)

SILICATE OF:	CHEMICAL NAME:	FORMULA:	MINERAL NAME:
Fe^{+2}	iron silicate	Fe_2SiO_4	fayalite
Mg^{+2}			forsterite
Fe^{+2}, Al^{+3}		$Fe_3Al_2(SiO_4)$	almandine
Ca^{+2}, Al^{+3}			grossular
Zr^{+4}	zirconium silicate		zircon

34

Introduction

Remind your students that these 8 basic elements comprise over 99% of all rocks and minerals. Explain how each element gains or loses electrons to form a stable outer shell.

Answers / Notes

2.

a. Because the tetrahedron has 1 silicon atom and 4 oxygen atoms.

oxygen: $(4)(-2) = -8$
silicon: $(1)(+4) = \underline{+4}$
           -4 (net charge)

b.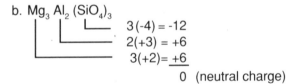

$Mg_3 \ Al_2 \ (SiO_4)_3$

$3(-4) = -12$
$2(+3) = +6$
$3(+2) = \underline{+6}$
       0 (neutral charge)

3a.

SILICATE OF:	CHEMICAL NAME:	FORMULA:	MINERAL NAME:
Fe^{+2}	iron silicate	Fe_2SiO_4	fayalite
Mg^{+2}	*magnesium silicate*	Mg_2SiO_4	forsterite
Fe^{+2}, Al^{+3}	*iron aluminum silicate*	$Fe_3Al_2(SiO_4)_3$	almandine
Ca^{+2}, Al^{+3}	*calcium aluminum silicate*	$Ca_3Al_2(SiO_4)_3$	grossular
Zr^{+4}	zirconium silicate	$ZrSiO_4$	zircon

3b. $Fe_2SiO_4 = (2)(2) + (1)(-4) = 0$
$Mg_2SiO_4 = (2)(2) + (1)(-4) = 0$
$Fe_3Al_2(SiO_4)_3 = (3)(2) + (2)(3) + (3)(-4) = 0$
$Ca_3Al_2(SiO_4)_3 = (3)(2) + (2)(3) + (3)(-4) = 0$
$ZrSiO_4 = (1)(4) + (1)(-4) = 0$

Materials

☐ Oil-based clay.

(TO) model how silicate tetrahedra join together in different ways, producing a variety of mineral forms.

HOW TETRAHEDRA COMBINE ○   Rocks and Minerals (   )

1. Make a circle with a penny. Project a *single* tetrahedron onto flat notebook paper like this.

2. Extend your single tetrahedron into a *double*. Make both a clay model and a penny projection.

CLAY MODEL

PENNY PROJECTION

shared atom

3. Join 6 tetrahedra in a *ring*. (Hint: Draw 6 circles around a center circle, then erase the center. Add 12 more oxygen plus 6 silicon.)

ERASE

4. Extend your ring into a *sheet*. Connect silicon atoms to form a repeating hexagon.

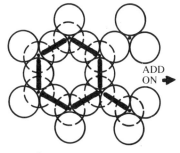

ADD ON ➡

5. Fold step 5 along any hexagon line to imagine how your sheet extends into a 3-dimensional *framework*.

FOLD

35

Introduction

The way silica tetrahedra fit together on a molecular scale is often repeated in a mineral's overall appearance. Quartz and feldspar, for example, form tough, hard 3-dimensional *frameworks* that fracture rather than cleave. Micas, by contrast, form flat hexagonal *sheets* that cleave in thin layers.

Answers / Notes

1. *As illustrated above.*

2. *As illustrated above. Notice that one oxygen atom is being shared by 2 tetrahedra. This makes a total of 7 oxygen plus 2 silicon.*

3.
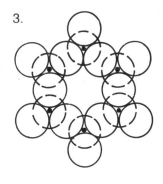

4.
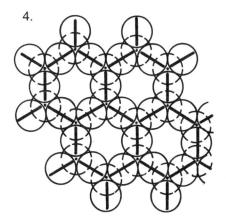

5. *As illustrated above. All 4 oxygen atoms of each tetrahedron are shared in this 3-D network, producing a neutral 1:2 ratio of silicon to oxygen (SiO_2) called silica. This is the only pure form of silicon and oxygen that forms a neutral mineral. All other forms of the silica tetrahedra (in steps 1-4) are negative, requiring additional positive ions from other elements to form neutral minerals.*

Materials
☐ A penny.
☐ The clay tetrahedra model from the previous task card.
☐ Oil based clay.

(TO) trace the path of a silica tetrahedron through the rock cycle.

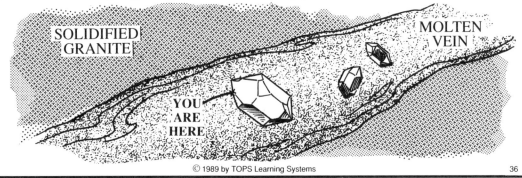

ADVENTURES OF TETRA   **Rocks and Minerals (   )**

You are a silica tetrahedron swimming in a vein of slowly cooling pegmatite. Many of your neighboring tetrahedra have solidified onto a large and growing crystal of pure quartz. Today it's your turn. You leave the turmoil of your magma melt to fit perfectly into its ordered array. Others follow and soon you find yourself locked deep inside that beautiful transparent crystal.

"How fortunate," you reflect. "To have a clear window on the world. Much better than basalt." Being part of a vein of glorious pegmatite, you anticipate the time, still far into the future, when erosion will uncover you and release you from the long night... (continue the story!)

SOLIDIFIED GRANITE

MOLTEN VEIN

YOU ARE HERE

36

Introduction

Help your students stay within the realm of science, not fiction, by discussing how silica tetrahedra combine with other elements to form minerals:
• In a magma melt, the tetrahedra ionize to stand alone as negative SiO_4^{-4}.
• In igneous quartz, sedimentary agate, metamorphic quartzite and other forms of crystallized silica, negative tetrahedra combine to form a neutral 3-dimensional framework (a 1:2 ratio of silicon to oxygen, SiO_2). Other atoms lock into this framework, providing a variety of colors and textures.
• In basalt, the negative tetrahedra associate with positive ions of magnesium and iron. These minerals weather into clay.
• In feldspar the tetrahedra form a neutral 3-dimension framework similar to quartz, with substitution of aluminum and potassium for some silicon. Feldspar also weathers into clay.

Answers / Notes

Encourage students to map out a story line before they begin to write. There are many possible directions. All eventually come full circle.

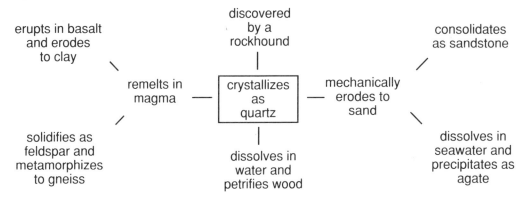

enrichment

REPRODUCIBLE
STUDENT
TASK CARDS

Task Cards Options

Here are 3 management options to consider before you photocopy:

1. Consumable Worksheets: Copy 1 complete set of task card pages. Cut out each card and fix it to a separate sheet of boldly lined paper. Duplicate a class set of each worksheet master you have made, 1 per student. Direct students to follow the task card instructions at the top of each page, then respond to questions in the lined space underneath.

2. Nonconsumable Reference Booklets: Copy and collate the 2-up task card pages in sequence. Make perhaps half as many sets as the students who will use them. Staple each set in the upper left corner, both front and back to prevent the outside pages from working loose. Tell students that these task card booklets are for reference only. They should use them as they would any textbook, responding to questions on their own papers, returning them unmarked and in good shape at the end of the module.

3. Nonconsumable Task Cards: Copy several sets of task card pages. Laminate them, if you wish, for extra durability, then cut out each card to display in your room. You might pin cards to bulletin boards; or punch out the holes and hang them from wall hooks (you can fashion hooks from paper clips and tape these to the wall); or fix cards to cereal boxes with paper fasteners, 4 to a box; or keep cards on designated reference tables. The important thing is to provide enough task card reference points about your classroom to avoid a jam of too many students at any one location. Two or 3 task card sets should accommodate everyone, since different students will use different cards at different times.

EIGHT BASIC ELEMENTS ○ Rocks and Minerals ( )

1. Most rocks and minerals are made from just 8 common elements. These are listed in bold in the periodic table below.

2. Spell out the names of these 8 elements.

PERIODIC TABLE OF THE ELEMENTS

1 H																	2 He
3 Li	4 Be	ATOMIC MASS — NUMBER								5 B	6 C	7 N	8 **O** 46.4%	9 F	10 Ne
11 **Na** 2.4%	12 **Mg** 2.3%			PERCENT TOTAL WEIGHT—						13 **Al** 8.2%	14 **Si** 28.2%	15 P	16 S	17 Cl	18 Ar
19 **K** 2.0%	20 **Ca** 4.2%	21 Sc	25 Mn	26 **Fe** 5.6%	27 Co	30 Zn	31 Ga	32 Ge	33 As	34 Se	35 Br	36 Kr

3. The percent (by weight) of each element is written under its symbol. Make a bar graph to show the relative abundance of each element. Add a 9th bar to represent all other elements.

USE LINED PAPER

1

TINY TREASURES ○ Rocks and Minerals ( )

1. Put about a spoonful of sand into a pie tin. Use a hand lens to inspect the tiny grains. Pick out different shapes, lusters and colors with the moistened tip of your pencil.

2. Copy this mineral table. For each box, find the correct grain sample, describe it, then tape it in.

CHUNKY TRANSLUCENT Quartz	description: *crystal clear* specimen:		
CHUNKY OPAQUE Quartz or Feldspar			
FLAT SHINY Mica			

3. Minerals *cleave* (break evenly) to form flat, smooth surfaces. Describe sand grains you see that have cleaved.

4. Minerals *fracture* (break unevenly) to form irregular, rough surfaces. Describe sand grains you see that have fractured.

2

ONE MINERAL / MANY FORMS ○   Rocks and Minerals (   )

1. Divide a small pinch of sugar into 3 tiny piles on a glass slide.

LARGE CRYSTALS
Leave this pile alone.

SMALL CRYSTALS
Grind to a fine powder with the back of your finger nail.

FUSED MASS
Heat gently until the sugar *just* melts.

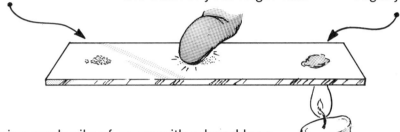

2. Examine each pile of sugar with a hand lens.
   a. Describe what you see.
   b. Propose a theory to explain why each pile of sugar looks the way it does.

3. A huge variety of different-looking rocks contain the same basic mix of just 8 elements. How does this experiment help explain such diversity?

3

IGNEOUS ROCK   ○   Rocks and Minerals (   )

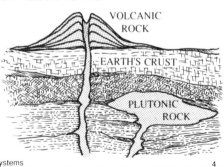

1. Lay down a row of saturated salt water drops across the diameter of an aluminum pie tin. Rest it on a jar so a candle fits under the overhang, directly below the first drop.
   a. Light the candle. Its flame should touch the bottom edge of the pie tin.
   b. Write your observations over the next 10 to 20 minutes.
   c. How is the size of a crystal related to how fast it forms?

2. Examine 2 samples of igneous rock — granite and basalt. Use a hand lens.
   a. Compare and contrast the crystallized minerals in each rock.
   b. Which rock is most likely *plutonic*? Most likely *volcanic*? Explain how you know.

4

MOH'S HARDNESS SCALE ○ Rocks and Minerals ( )

1. Try to scratch a penny with a paper clip. . .   . . .Try to scratch a paper clip with a penny. . .   . . .Which is harder and which is softer? Why?

RUB HARD

HARDER?
SOFTER?

2. These minerals define Moh's scale of hardness, where 1 is the softest and 10 is the hardest. What scratches what on this scale? Explain your reasoning.

soft ←   → hard

1   gypsum   3   flourite   5   orthoclase   7   topaz   9   diamond

talc   2   calcite   4   apatite   6   quartz   8   corundum   10

3. Brick is rated about 4.5 on Moh's hardness scale.
   a. Predict what happens if you rub the mineral quartz across brick?
   b. Would you get the same result with calcite? Explain.

4. Place these 6 items on a hardness scale, along with brick, in the correct order: penny, chalk, glass jar, fingernail, paper clip.

                              (4.5) brick

soft ←————————————————————+————————————————————→ hard

5

STREAK TEST ○ Rocks and Minerals ( )

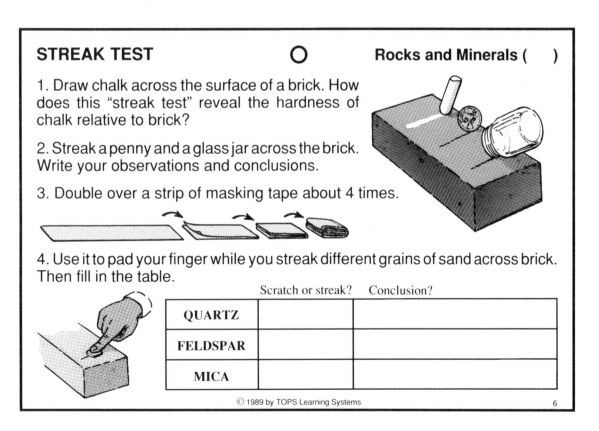

1. Draw chalk across the surface of a brick. How does this "streak test" reveal the hardness of chalk relative to brick?

2. Streak a penny and a glass jar across the brick. Write your observations and conclusions.

3. Double over a strip of masking tape about 4 times.

4. Use it to pad your finger while you streak different grains of sand across brick. Then fill in the table.

	Scratch or streak?	Conclusion?
QUARTZ		
FELDSPAR		
MICA		

6

MECHANICAL WEATHERING   Rocks and Minerals (   )

Grind 2 granite rocks together as hard as you can for about 1 minute. Collect the falling fragments on clean notebook paper, then tap them together into one central pile.

1. Describe the tiny fragments that eroded from the parent rock. Can you still recognize them as crystals?

2. How has the texture of the parent rock changed? Propose a theory to explain why rocks look smooth and dull on the outside.

3. How do rocks mechanically weather in nature? List as many natural erosion forces as you can.

4. Where does topsoil come from? Why is it important to conserve topsoil?

7

TRANSPORT AND SORTING O   Rocks and Minerals (   )

1. Grind 2 granite rocks together for about 1 minute, as before. Collect the *clasts* (rock fragments) on clean notebook paper.

2. Hold the paper like a funnel and pour the rock fragments into a test tube. Notice how gravity sorts these clasts by size as they move down the paper.
 a. Write your observations.
 b. Explain how wind has a similar sorting effect.

CLASTS
SORTED BY
GRAVITY

3. Add a few drops of water to the test tube and shake it up.
 a. Write your observations.
 b. Discuss the role that rivers play in transporting and sorting sediment.

4. Add a centimeter of sand to 2 test tubes. Fill one with water; leave the other filled with air. Then vigorously shake both test tubes.
 a. After 1 minute, make labeled drawings of each one.
 b. Contrast the *sorted* sediment with *unsorted* sediment.

WATER   AIR
1 cm

8

SEDIMENTARY ROCK ⭕ Rocks and Minerals ( )

1. Grind 2 granite rocks together for about 1 minute as before. Collect the clastic debris on clean notebook paper.

2. Transfer the sediment to one end of a microscope slide. Add just *one* drop of dilute hydrochloric acid and mix with a paper clip to form a tiny mud puddle.

3. Gently heat to dryness over a candle flame, then allow the slide to cool.

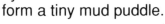

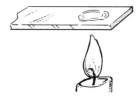

4. Try to blow the dried mud off the slide. Notice that you have begun to form a new rock.
   a. Why is "clastic sedimentary rock" a good name for this new material?
   b. Compare the texture of this new rock with its parent rock.

5. Dissolved minerals in the acid drop *cement* the clasts in your new rock together. Brush the slide lightly with your fingertips to discover where this cementation is the strongest.

9

CHEMICAL WEATHERING ⭕ Rocks and Minerals ( )

1. Stand a short piece of blackboard chalk on a flat lid.

2. Surround it with about 4 eye-droppers full of 5% hydrochloric acid.

3. Calcium carbonate (chalk) dissolves in acid as it produces carbon dioxide gas. What evidence for this reaction do you *see* and *hear*?

4. Examine the sides of the chalk above the water line with a hand lens.
   a. Draw the tiny drainage patterns that open up as the calcium carbonate dissolves away.
   b. Whole mountains are made from calcium carbonate (limestone). Draw a picture to show chemical weathering on this grander scale.

5. How is this chemical weathering of chalk different from the mechanical weathering of granite you studied earlier?

10

SEA FLOOR SEDIMENT   O   Rocks and Minerals (   )

1. Examine the chalk that was weathered by 5% hydrochloric acid.

    a. Draw and label a "before" and "after" picture.

    b. Account for the changes in each picture.

BEFORE   AFTER

2. Put a drop of 5% hydrochloric acid on a sea shell.

    a. Write your observations.

    b. What are shells made from? Where do you think sea animals might get this compound to make their shells?

    c. What happens to their shelled remains after these animals die?

3. Blackboard chalk is an accumulation of the shells of microscopic sea animals. Show by diagram how the chalk in your classroom became a chemical sedimentary rock.

11

TWO KINDS OF CEMENT   O   Rocks and Minerals (   )

1. Pick out a large transparent quartz crystal from some sand. Fold it inside a small square of paper, then crush it with a pair of pliers.

2. Gather the silica residue in the left corner of a glass slide. Put an equal-sized lump of blackboard chalk in the right corner.

3. Put 1 drop of dilute hydrochloric acid over each mineral. Keep the drops separate.

SILICA   CHALK

4. Heat to dryness very slowly, moving the slide back and forth over a candle flame. (It splatters if you heat too quickly.) Then let it cool.

5. Draw a picture of the silica residue, labeling the quartz clasts and silica cement. Cite examples of rocks that contain these ingredients.

6. Draw a picture of the chalk residue, labeling the calcite clasts and calcium carbonate cement. Cite examples of rocks that contain these ingredients.

12

CRACKING UP ○ Rocks and Minerals ( )

1. Tape your name on a large tin can. Use it to store the freshly broken rock samples you will collect in step 2.

2. Crack open small rocks between large rocks. CAUTION: Always protect your eyes from sharp flying fragments:

   a. Wear safety goggles,

   (or)

   b. Hold the target rock inside a plastic bag before you strike it.

3. What to keep:
   • Small pieces, about as big as your thumb to the first joint.
   • All kinds of textures and colors.
   • One-of-a-kind pieces. Discard or trade duplicates.

13

ROCK HARDNESS ○ Rocks and Minerals ( )

1. You have streak tested minerals. Here's how to streak test a rock.

a. Know the difference between streaking and crumbling…

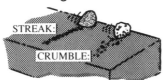

What kind of rock might crumble against brick yet still scratch?

b. Rub surface to surface. Don't gouge with a point…

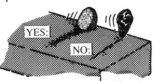

Make a paper clip (4.0) both streak and scratch a brick (4.5). Explain which is the valid test.

c. Test only freshly broken surfaces…

YES: FRESH   NO: WEATHERED
SURFACE   SURFACE

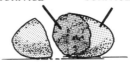

Streaking the outside of a weathered rock might give a false test. Why is this so?

2. Test each rock in your collection can for hardness:
• Tag those that *streak* "st."
• Tag those that *crumble and scratch* "cr."
• Don't tag those that *scratch*.

14

METAMORPHIC ROCK   Rocks and Minerals (   )

1. Intense heat and pressure transforms igneous or sedimentary rock into *metamorphic* (changed) rock. Where on earth might rocks be exposed to such conditions?

2. Sort the rocks in your collection can into these 2 groups:

FOLIATED
ordered repeating texture

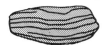

MASSIVE
random nonrepeating texture

3. a. Look for these metamorphic rocks in your *foliated* group.

> Slate: has hard thin layers; comes from shale.

> Schist: is banded by mica; comes from granite.

b. Look for these metamorphic rocks in your *massive* group.

> Quartzite: has a sugary luster; comes from sandstone.

> Marble: has large white crystals; comes from limestone.

4. Describe in words and pictures each type of rock you are able to classify. (If you find none, describe a friend's example.)

15

SORTING ROCKS   Rocks and Minerals (   )

1. Cut notebook paper to size so it just fits inside an egg-carton lid. Divide it into 12 equal squares.

2. Neatly copy all of this information into the squares, then tape them inside the lid. Number the egg cups in the same order.

MAY REQUIRE SPLITTING

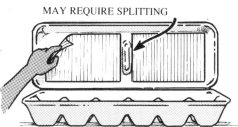

1	2	3	4	5	6
CLASTIC SEDIMENTARY coarse grained	CLASTIC SEDIMENTARY fine grained	CLASTIC SEDIMENTARY poorly sorted	CHEMICAL SEDIMENTARY carbonate	CHEMICAL SEDIMENTARY silica	CHEMICAL SEDIMENTARY other
7	**8**	**9**	**10**	**11**	**12**
PLUTONIC IGNEOUS coarse crystals	VOLCANIC IGNEOUS fine crystals	VOLCANIC IGNEOUS pyroclastic	MASSIVE METAMORPHIC quartzite/marble	FOLIATED METAMORPHIC schist/gneiss	METAMORPHIC slate/other

3. Sort rocks from your can into the correct egg cups.
   • If you have no idea how to classify a rock, leave it in your can.
   • If you have some idea where a rock might belong, go ahead and place it where you think it goes. You can always change your mind later.

16

IDENTIFYING ROCKS ○ Rocks and Minerals ( )

Use the next 4 task cards (in any order) to help identify rocks in your collection. When you are sure about a rock, fill out an ID card like this one, using quarter pieces of notebook paper...

3. clastic sedimentary, poorly sorted: CONCRETE

gravel clasts
sand clasts
carbonate cement matrix

TESTS: Fizzes with dilute acid; crumbles across brick.

DESCRIPTION: A conglomerate of rounded sand and gravel clasts cemented in a carbonate matrix.

FORMATION: A human-made rock. Probably once part of a side-walk or building foundation.

◀ 1. egg cup category
◀ 2. name
◀ 3. labeled drawing
◀ 4. tests
◀ 5. description
◀ 6. formation

...Continue in this manner until you classify from 1 to 3 rocks per egg cup, at least 24 rocks in all. This will require several days of thoughtful work. During this time keep looking for rocks to fill empty positions, or trade with friends.

17

CLASTIC SEDIMENTARY (1-3) ○ Rocks and Minerals ( )

1	2	3
coarse grained	**fine grained**	**poorly sorted**

1 — coarse grained

SANDSTONE

coarse   medium

always:
• clasts have uniform size (well sorted)
• clasts are big enough to see individually
usually:
• rough to touch (unless worn smooth)
• cemented by silica
sometimes:
• crumbles across brick
• scratches brick
occasionally:
• contains fossils, water ripples
• has bedded layers

2 — fine grained

CLAY OR SILTSTONE

SHALE

always:
• smooth to touch
• clasts too small to see individually
usually:
• cemented by clay
• streaks across brick
• breaks along flat horizontal surfaces
sometimes:
• has bedded layers
occasionally:
• contains fossils, water ripples

3 — poorly sorted

CONGLOMERATE

ORE BEARING

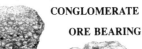

always:
• poorly sorted
usually:
• large and small clasts (poorly sorted sizes)
• color variations (poorly sorted minerals)
• crumbles or scratches across brick
sometimes:
• cemented by carbonates, silica, or clay; if rust colored, probably cemented by iron
• bubbles with dilute acid

18

CHEMICAL SEDIMENTARY (4-6) ○   Rocks and Minerals (   )

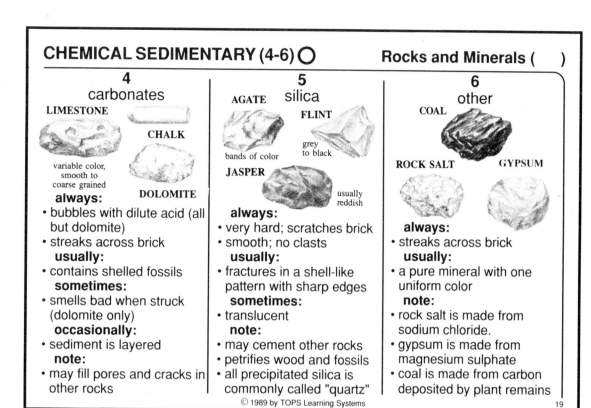

4
carbonates

LIMESTONE

CHALK

variable color, smooth to coarse grained

DOLOMITE

always:
- bubbles with dilute acid (all but dolomite)
- streaks across brick

usually:
- contains shelled fossils

sometimes:
- smells bad when struck (dolomite only)

occasionally:
- sediment is layered

note:
- may fill pores and cracks in other rocks

5
silica

AGATE

FLINT

grey to black

bands of color

JASPER

usually reddish

always:
- very hard; scratches brick
- smooth; no clasts

usually:
- fractures in a shell-like pattern with sharp edges

sometimes:
- translucent

note:
- may cement other rocks
- petrifies wood and fossils
- all precipitated silica is commonly called "quartz"

6
other

COAL

ROCK SALT

GYPSUM

always:
- streaks across brick

usually:
- a pure mineral with one uniform color

note:
- rock salt is made from sodium chloride.
- gypsum is made from magnesium sulphate
- coal is made from carbon deposited by plant remains

© 1989 by TOPS Learning Systems   19

IGNEOUS (7-9)   ○   Rocks and Minerals (   )

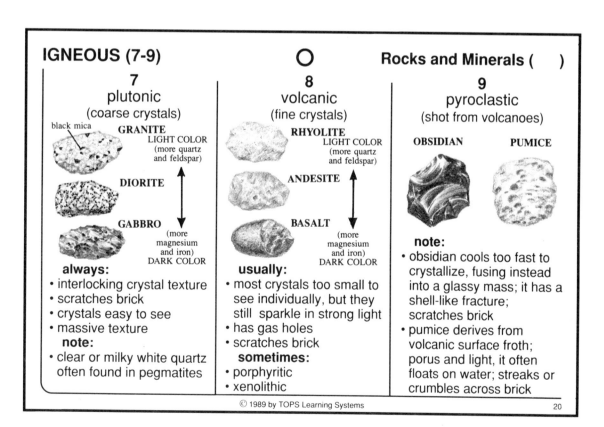

7
plutonic
(coarse crystals)

black mica

GRANITE

LIGHT COLOR (more quartz and feldspar)

DIORITE

GABBRO

(more magnesium and iron) DARK COLOR

always:
- interlocking crystal texture
- scratches brick
- crystals easy to see
- massive texture

note:
- clear or milky white quartz often found in pegmatites

8
volcanic
(fine crystals)

RHYOLITE

LIGHT COLOR (more quartz and feldspar)

ANDESITE

BASALT

(more magnesium and iron) DARK COLOR

usually:
- most crystals too small to see individually, but they still sparkle in strong light
- has gas holes
- scratches brick

sometimes:
- porphyritic
- xenolithic

9
pyroclastic
(shot from volcanoes)

OBSIDIAN

PUMICE

note:
- obsidian cools too fast to crystallize, fusing instead into a glassy mass; it has a shell-like fracture; scratches brick
- pumice derives from volcanic surface froth; porus and light, it often floats on water; streaks or crumbles across brick

© 1989 by TOPS Learning Systems   20

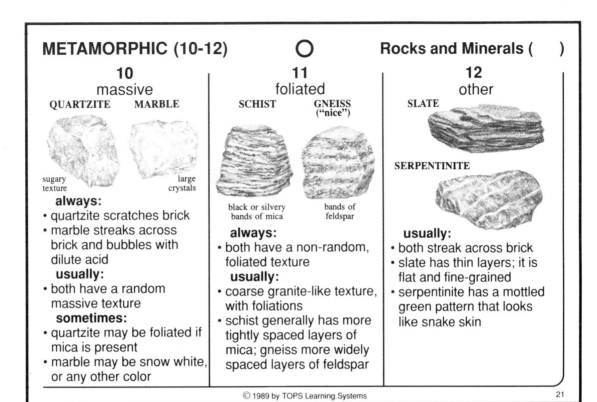

METAMORPHIC (10-12)   O   **Rocks and Minerals (   )**

10
massive
QUARTZITE   **MARBLE**

sugary texture
large crystals

always:
- quartzite scratches brick
- marble streaks across brick and bubbles with dilute acid

usually:
- both have a random massive texture

sometimes:
- quartzite may be foliated if mica is present
- marble may be snow white, or any other color

11
foliated
SCHIST   **GNEISS ("nice")**

black or silvery bands of mica
bands of feldspar

always:
- both have a non-random, foliated texture

usually:
- coarse granite-like texture, with foliations
- schist generally has more tightly spaced layers of mica; gneiss more widely spaced layers of feldspar

12
other
SLATE

SERPENTINITE

usually:
- both streak across brick
- slate has thin layers; it is flat and fine-grained
- serpentinite has a mottled green pattern that looks like snake skin

21

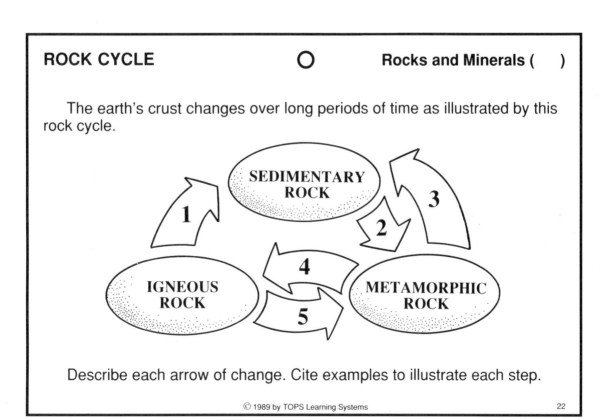

ROCK CYCLE   O   **Rocks and Minerals (   )**

The earth's crust changes over long periods of time as illustrated by this rock cycle.

SEDIMENTARY ROCK

1

2   3

IGNEOUS ROCK

4

5

METAMORPHIC ROCK

Describe each arrow of change. Cite examples to illustrate each step.

22

ORDINARY ROCK   O   Rocks and Minerals ( )

Over 99% of the earth's crust is made from the following rocks:

dark igneous (mostly basalt) = 42.7%   marble = 0.9%
gneiss = 21.4%   sandstone = 1.7%
light igneous (mostly granite) = 22.0%   schist = 5.1%
limestone = 2.0%   shale = 4.2%

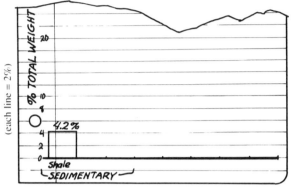

1. What percent of the earth's crust is made from…
   a. sedimentary rock?
   b. metamorphic rock?
   c. igneous rock?

2. Represent all 8 ordinary rocks on a bar graph. Arrange them in this order: sedimentary – metamorphic – igneous.

3. Some of these rocks don't seem to be all that common; at least they were hard to find for your rock collection. Explain this discrepancy.

23

PERMEABILITY   O   Rocks and Minerals ( )

1. Get a piece of chalk and some oil-based clay about as big as your thumb. Find the mass of each to the nearest tenth of a gram.

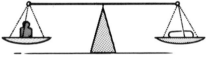

2. Touch a single drop of water to each substance.
   a. Which material has the greatest permeability (allows water to soak through most easily)?
   b. Predict which material will gain the most mass if you soak both in water.

CLAY   CHALK

3. Soak the chalk and clay in a jar of water for about 1 minute.
   a. Write your observations.
   b. Dry each substance with a paper towel, then reweigh to the nearest tenth of a gram. Evaluate your prediction.

4. Find the most permeable rock in your collection. What kind of rock is it? Does your answer make sense? Explain.

24

DENSITY **Rocks and Minerals (   )**

1. Fill a graduated cylinder with exactly 50.0 ml of water. Add clay pebbles until the water level raises to 60.0 ml. If you overshoot, try again.

   a. What volume of clay is in the cylinder?

   b. Remove the clay from the water, dry the pebbles and press them together. Find the mass of this lump to the nearest tenth gram.

2. Density is defined as the mass of any substance divided by its volume.

$$D = \frac{\text{mass}}{\text{volume}} = \frac{g}{ml}$$

   a. Find the density of your clay ball.

   b. In a similar manner, make a 20 ml ball of clay and find its density.

   c. Make a 5 ml ball of clay and find its density.

3. How do the densities of each clay ball compare? Make a generalization about density.

25

DENSITY OF IGNEOUS ROCK ○   **Rocks and Minerals (   )**

1. Cut a circle of paper that *just* fits inside your graduated cylinder.

2. Use this circle as a template to find 4 pieces of granite that are reasonably large, but still fit inside your cylinder.

   a. Weigh these 4 pieces of granite together, to the nearest tenth gram.

   b. Find the volume of all 4 pieces in your graduated cylinder.

   c. Divide mass by volume to find the density of granite.

3. In a similar manner, find the density of four pieces of basalt.

4. Ocean basins are composed mainly of basalt; the continents of granite and related rocks. Use your density results to propose a theory that might explain why this is so.

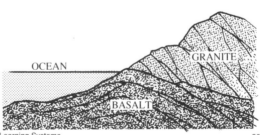

OCEAN   GRANITE   BASALT

26

DENSITY OF SILICA
○
Rocks and Minerals (   )

1. Use a penny to draw circles on paper . . .

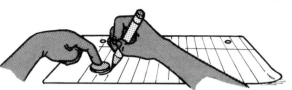

a. Arrange 15 circles into a tight, organized pack.

b. Arrange another 15 circles into a loose, disorganized pack.

MINIMUM SPACE

RANDOM SPACE

2. The silica in agates crystallizes in a slow, organized way. The silica in glass marbles fuses quickly into a disorganized mass.

    a. How would you expect the density of glass marbles to compare with the density of agates? Use your drawings in step 1 to support your hypothesis.
    b. Test your hypothesis. Take measurements and calculate densities.

27

SPECIFIC GRAVITY
○
Rocks and Minerals (   )

1. Find the density of 100 ml of water. Show your work.
    a. Check your answer by finding the density of 50 ml of water.
    b. The density of water is always very close to what whole number?

2. *Specific gravity* is the density of any substance divided by the density of water.

$$S.G. = \frac{\text{density of substance}}{\text{density of water}}$$

    a. What is the specific gravity of granite? Of basalt?
    b. How does the specific gravity of a substance compare with its density?

3. Pumice often floats on water. What can you conclude about the specific gravity of this rock?

4. The whole earth, on average, is 5.5 times denser than water. (It has a specific gravity of 5.5.)
    a. Compare this with the specific gravity of rocks in the earth's crust.
    b. How is matter near the center of the earth different from matter near its surface?

28

ICE AND WATER Rocks and Minerals (   )

1. Gently dunk a film canister in a tub of standing water without stirring it up. Snap on the cap while still under water so no air pockets remain inside.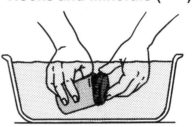

2. Freeze it overnight.

   a. As ice freezes, does its volume increase or decrease? How do you know?

   b. Predict what happens to the film canister after the water freezes.

   c. Wait 24 hours to complete the rest of this task card.

3. Measure the ice and canister with a mm ruler. Calculate the % increase in the length of the ice.

$$\% \text{ increase} = \frac{\text{extra length}}{\text{original length}} \times 100$$

4. Slide the ice out of its canister. Draw a diagram to show where dissolved air and ice cracks form in the ice.

5. Freezing and melting water is the main cause of mechanical rock erosion. Explain how this works.

29

CONCENTRATION AND CRYSTAL SIZE Rocks and Minerals (   )

1. Divide a lid or glass pane into 4 parts. Label them with these letters:

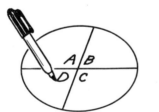

2. Add the given number of saturated *salt water* drops to each section with an eyedropper.

3. Now dilute these salty drops by adding the given number of *fresh water* drops to each section.

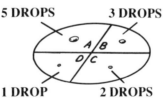

5 DROPS	3 DROPS	NO DROPS	3 DROPS
A	B	A	B
D	C	D	C
1 DROP	2 DROPS	7 DROPS	6 DROPS

4. Find the % of salt water saturation in each puddle:

$$\% = \frac{\text{salt water drops}}{\text{total drops}} \times 100$$

5. Predict the relative sizes of salt crystals that remain after the water evaporates from each puddle. Explain your reasoning.

6. Evaluate your prediction after 24 hours. Report, in words and pictures, how mineral concentration relates to crystal size.

30

PEGMATITES Rocks and Minerals (   )

A freezing film canister of water closely models a cooling pluton of granite. Study these similarities, then answer the 3 questions below.

FREEZING WATER CANISTER
(1/2 actual size)

Crack rich in dissolved air — Cold air
Crystallized water (ice)
Liquid water — Concentrated air

COOLING GRANITE PLUTON
(greatly reduced)

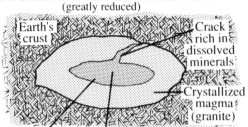

Earth's crust — Crack rich in dissolved minerals
Crystallized magma (granite)
Liquid magma — Concentrated minerals

1. Dissolved air doesn't easily freeze in water. How does this fact explain why air concentrates in the center of the ice, as it did in activity 29?

2. Water vapor, quartz, mica, feldspar, tourmaline, beryl and topaz all tend to concentrate in the center of cooling granite plutons. What do these minerals share in common with dissolved air?

3. Granite cracks as it solidifies and cools. Use results from activity 30 to explain why these cracks fill with extra large crystals, called pegmatites.

31

PETRIFIED PAPER Rocks and Minerals (   )

1. Fill a small jar half full with Epsom salts (magnesium sulfate crystals). Fill with water and stir well.

2. Bend a paper clip and tape it to the jar like this. Set the jar on a wide plastic lid.

3. Cut a paper towel in 2 halves. Crumple one into a ball; fold the other into a narrow strip. Soak both in the Epsom salts.

SOAKED TOWELS

4. Put the strip in the jar with one end resting on the paper clip. Rest the crumpled paper on the jar's rim. Set this experiment aside to use in the next activity.

5. Wood petrifies (turns to stone) as dissolved silica replaces wood fiber. Predict how magnesium sulfate might "petrify" this paper towel in a similar manner.

32

1. Set your "petrifying" paper in a warm dry place. *After* it is totally dry, answer these questions.

   a. How did the paper change?

   b. Will it burn in a candle flame?

   c. How is this paper similar to petrified wood? Evaluate your prediction.

2. Adjust the paper strip so the end hangs over the paper clip, slightly below the level of solution in the jar. It should drip *very* slowly.

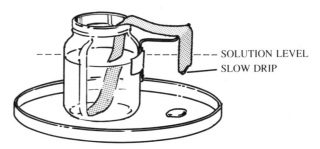

   — SOLUTION LEVEL
   — SLOW DRIP

a. How do stalactites and stalagmites form in caves? What are they made from? (Use a dictionary if you need one.)

b. See if you can grow these formations from magnesium sulfate over the next few days. Report your results in words and pictures.

---

1. If oxygen is drawn as large as a penny, here's how the 8 basic elements compare as ions:

O^{-2} K^{+1} Ca^{+2} Na^{+1} Mg^{+2} Fe^{+2} Al^{+3} Si^{+4}

2. Roll 4 clay "oxygen" balls to the diameter of a penny. Fit these around 1 tiny "silicon" ball to model a *silica tetrahedron*.

   a. Why do chemists call these atoms "SiO_4"? Add up their net charge.

   b. Minerals that contain these tetrahedra are called "silicates". They are always neutral. Show that $Mg_3Al_2(SiO_4)_3$ is a neutral silicate.

3. Silicates comprise 92% of all rock-forming minerals!

   a. Complete this table.
   b. Show that each silicate is neutral.

   (Save your model.)

SILICATE OF:	CHEMICAL NAME:	FORMULA:	MINERAL NAME:
Fe^{+2}	iron silicate	Fe_2SiO_4	fayalite
Mg^{+2}			forsterite
Fe^{+2}, Al^{+3}		$Fe_3Al_2(SiO_4)$	almandine
Ca^{+2}, Al^{+3}			grossular
Zr^{+4}	zirconium silicate		zircon

HOW TETRAHEDRA COMBINE ◯   Rocks and Minerals ( )

1. Make a circle with a penny. Project a *single* tetrahedron onto flat notebook paper like this.

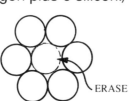

2. Extend your single tetrahedron into a *double*. Make both a clay model and a penny projection.

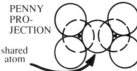

CLAY MODEL

PENNY PRO- JECTION

shared atom

3. Join 6 tetrahedra in a *ring*. (Hint: Draw 6 circles around a center circle, then erase the center. Add 12 more oxygen plus 6 silicon.)

ERASE

4. Extend your ring into a *sheet*. Connect silicon atoms to form a repeating hexagon.

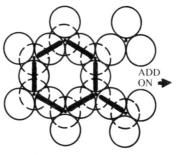

ADD ON ➡

5. Fold step 5 along any hexagon line to imagine how your sheet extends into a 3-dimensional *framework*.

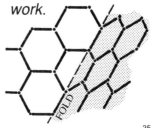

FOLD

© 1989 by TOPS Learning Systems   35

ADVENTURES OF TETRA   ◯   Rocks and Minerals ( )

You are a silica tetrahedron swimming in a vein of slowly cooling pegmatite. Many of your neighboring tetrahedra have solidified onto a large and growing crystal of pure quartz. Today it's your turn. You leave the turmoil of your magma melt to fit perfectly into its ordered array. Others follow and soon you find yourself locked deep inside that beautiful transparent crystal.

"How fortunate," you reflect. "To have a clear window on the world. Much better than basalt." Being part of a vein of glorious pegmatite, you anticipate the time, still far into the future, when erosion will uncover you and release you from the long night… (continue the story!)

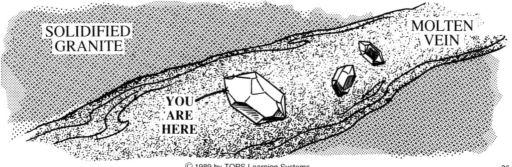

SOLIDIFIED GRANITE

MOLTEN VEIN

YOU ARE HERE

© 1989 by TOPS Learning Systems   36